Agroecology

Agroecology

The Scientific Basis of Alternative Agriculture

Miguel A. Altieri

with contributions by
Richard B. Norgaard, Susanna B. Hecht,
John G. Farrell, and Matt Liebman

Westview Press (Boulder)
IT Publications (London)
1987

This book is included in Westview's Special Studies in Agriculture Science and Policy.

All rights reserved. No part of this publication may be reproduced or transmitted in any form or by any means, electronic or mechanical, including photocopy, recording, or any information storage and retrieval system, without permission in writing from Westview Press.

Copyright © 1987 by Miguel A. Altieri

Published in 1987 in the United States of America by Westview Press, Inc.; Frederick A. Praeger, Publisher; 5500 Central Avenue, Boulder, Colorado 80301

Published in 1987 in the United Kingdom by Intermediate Technology Publications, 9 King Street, London WC2E 8HW, England

Library of Congress Cataloging-in-Publication Data
Altieri, Miguel A.
  Agroecology: the scientific basis of alternative agriculture.
  Bibliography: p.
  Includes index.
  1. Agricultural ecology. I. Title.
S589.7.A47 1987   630'.2'745   86-15756
ISBN (U.S.) 0-8133-7284-4

ISBN (U.K.) 1-85339-000-3

Composition for this book was provided by the author.
This book was produced without formal editing by the publisher.

Printed in the United States of America

 The paper used in this publication meets the requirements of the American National Standard for Permanence of Paper for Printed Library Materials Z39.48-1984.

6 5 4 3 2 1

Contents

List of Tables and Figures ix
Preface xiii
Acknowledgments xvii

PART ONE
THE THEORETICAL BASIS OF AGRICULTURAL ECOLOGY

1 THE EVOLUTION OF AGROECOLOGICAL THOUGHT 1
   Susanna B. Hecht

   Historical Background 1
   What is Agroecology? 4
   Influences on Agroecological Thought 8

2 THE EPISTEMOLOGICAL BASIS OF
   AGROECOLOGY 21
   Richard B. Norgaard

   Characteristics of Western Thought 22
   The Agroecological World View 23
   Blending Western and "Traditional" Knowledge 25
   Notes 27

3 THE AGROECOSYSTEM: DETERMINANTS, RESOURCES
   AND PROCESSES 29

   Classification of Agroecosystems 29
   The Resources of an Agroecosystem 31
   Ecological Processes in the Agroecosystem. 33
   The Stability of Agroecosystems 39

PART TWO
THE DESIGN OF ALTERNATIVE AGRICULTURAL SYSTEMS AND TECHNOLOGIES

4 GENERATING SUSTAINABLE TECHNOLOGIES 47

Consequences of Inappropriate Technology 47
Alternative Production Systems 48
Farming Systems Research 50
Technology Research 56

5 DESIGNING SUSTAINABLE AGROECOSYSTEMS 59

Choosing an Agricultural System 59
Choosing a Cropping System 64
Crop Characteristics and Cropping Patterns 65

6 TRADITIONAL AGRICULTURE 69

Ecological Features of Traditional Agriculture 69
Advantages of Crop Diversity 73
Examples of Traditional Farming Systems 75
Traditional Ethnobotanical Knowledge 88
Traditional Agroecosystems and Genetic Resources 89
Notes 91

7 ECOLOGICALLY BASED AGRICULTURAL DEVELOPMENT PROGRAMS 93

Combining Traditional and Modern Technologies 93
Examples of Programs 94
Implications for the Future 104

8 ORGANIC FARMING IN NORTH AMERICA 107

Characteristics of Organic Farming 107
Constraints to Organic Farming 113

PART THREE
ALTERNATIVE PRODUCTION SYSTEMS

9 POLYCULTURE CROPPING SYSTEMS 115
Matt Liebman

The Prevalence of Polycultures Around the World 115

Yield Advantages 116
Effects of Polycultures on Insect Pests, Diseases,
Nematodes and Weeds 121
Future Directions 125
Notes 125

10 COVER CROPPING AND MULCHING 127

Benefits of Cover Cropping 127
Types of Cover Crop Management 129
Living Mulches 132
Cropping Systems with Legume Cover Crops 133

11 CROP ROTATION AND MINIMUM TILLAGE 139

Benefits of Rotating Crops 139
Minimum Tillage Systems 142

12 AGROFORESTRY SYSTEMS 149
John G. Farrell

Characteristics of Agroforestry 149
Classification of Agroforestry Systems 150
Designing Agroforestry Systems 155
Notes 158

PART FOUR
ECOLOGICAL MANAGEMENT OF INSECT PESTS,
PATHOGENS AND WEEDS

13 PEST MANAGEMENT 159

Agroecology and Pest Management 159
The Importance of Diversity 164

14 WEED ECOLOGY 173

Crop/Weed Competition 174
Allelopathy 176
Weed Management 178

15 PLANT DISEASE ECOLOGY AND MANAGEMENT 187

The Disease Triangle 187
Biological Control of Plant Pathogens 190

PART FIVE
TOWARD SUSTAINABLE AGRICULTURE

16 TOWARD SUSTAINABLE AGRICULTURE 195

    The Problems of Modern Agriculture 195
    The Transition 197

BIBLIOGRAPHY 201
ABOUT THE CONTRIBUTORS 223
INDEX 225

Tables and Figures

TABLES

1.1 The contrast in physical and socio-economic conditions of resource-rich vs. resource-poor farmers . 19

3.1 Agroecosystem determinants that determine the type of agriculture in each region 30

3.2 Factors influencing agricultural intensification in African regions where shifting cultivation is practiced 32

3.3 Structural and functional differences between natural ecosystems and agroecosystems 40

4.1 Strategies to augment yield of individual crops 49

4.2 Some agricultural technology approaches to reduce energy inputs into food production systems 51

4.3 A comparison between a typical top-down technology transfer approach (TOT) and a farming systems research approach (FSR) involving the farmers 54

4.4 A checklist of information on crop management practices to be recorded in each farm throughout the year 55

5.1 Factors affecting the choice of farming systems . . . 60

5.2	Some factors favoring success in modern agroecosystems	61
6.1	Some examples of soil, space, water and vegetation management systems used by traditional agriculturalists throughout the world	72
6.2	Agroclimatic crop zones of the central Andes	83
9.1	Polyculture land use efficiencies expressed as land equivalent ratios	117
10.1	Partial list of species and some management characteristics of cover crop plants recommended for California orchards and vineyards	131
10.2	Growth characteristics of legume species used as cover crops	133
10.3	Effect of overseeding legume cover crops on corn yield and weed stand, 1981	134
11.1	Effect of crop rotation of corn on insect populations or potential damage	140
11.2	Intensity and efficiency of energy use in continuous corn compared with crop rotations incorporating grain and forage legumes	142
13.1	Vegetational arrangement of crops and other plants in time and space evaluated as to the potential development of pest problems	162
13.2	Possible effects of intercropping on insect pest populations	165
13.3	Selected examples of multiple cropping systems that effectively prevent insect-pest outbreaks	168
13.4	Commercial root yield of cassava in monoculture and in polyculture with cowpea, with and without insecticide applications	170
14.1	A list of non-chemical methods of managing weeds and the ecological principles upon which each is based	180

14.2	Selected examples of cropping systems in which the presence of weeds enhanced the biological control of specific crop pests	183
15.1	Some features of the crop habitat influencing the spread of crop diseases	188
15.2	General methods of disease control and their epidemiologic effects	189
15.3	Economic, social, biological and environmental factors affecting prospects for cultural control of crop diseases	191
15.4	Dry and decayed soil amendments that reduce some diseases caused by soil-borne fungi	192
15.5	Examples of green manures which reduce some soil-borne fungal pathogens	193
15.6	Soil amendments found to reduce nematode populations	193

FIGURES

2.1	The dominant, or mechanical world view	23
2.2	The coevolution of knowledge, values, social organization, technology, and the biological system	24
3.1	The general structure of an agricultural system and its relationship with external systems	30
3.2	Inputs, outputs and energy ratios of seven agricultural systems.	35
3.3	Five possible cropping systems that fit a rainfall pattern in Southeast Asia	38
3.4	Ecological patterns of contrasting agroecosystems	41
3.5	Degree of environmental control necessary for the maintenance of normal levels of productivity in three types of farming systems	42

3.6	The system properties of agroecosystems and indices of performance	43
4.1	Methodological sequence in the modification of farmer's agroecosystem management plan	58
5.1	Relationship between land capability classification classes and the intensity with which each class can be used	63
5.2	Climatic diagram of the Central Plateau of Mexico	66
6.1	Scheme of a small farming system with four production/consumption systems	70
6.2	Conceptual model of the production system of a Nepalese hill farm	70
6.3	Structural layout of a small-scale intensive farming system in the coastal zone of central Chile	85
6.4	Structural layout of a twelve-hectare extensive semi-commercial farming system in southern Chile	87
7.1	Model design of a self-sufficient farming system based on a seven-year rotational scheme adaptable to Mediterranean environments	98
7.2	Diagrammatic representation of a modular system emphasizing balancing inputs and outputs by various ecological management practices	102
7.3	The potential performance of Green Revolution technologies (high-input agriculture) and agroecological technologies (low-input agriculture) along a gradient of natural resource and socio-economic conditions affecting peasant farming systems.	105
12.1	The influence of trees in Tlaxcala, Mexico, on the growing environment of maize	152
12.2	Change in surface soil nitrogen and phosphorus with increasing distance from individual capulin and sabino trees	153

Preface

There are at least two types of farmers who do not approach agriculture in a conventional way. These are the small farmers of underdeveloped countries (especially the tropics) and an emerging group of organic farmers in Europe and the United States. Through their practices, these farmers have at times inadvertently challenged chemical and mechanized agriculture. The conventional view of these farming systems by western agriculturalists is that they are "backward" and constitute a use of archaic and outmoded technologies because they have evolved through trial and error, outside the mainstream of scientific research and agricultural extension programs. Therefore, scientific experts, industrialists and government officials are quick to discredit these systems because their methods of production have not been "scientifically" proved.

Unfortunately, the formal analysis of these systems is sometimes manipulated to deliberately discredit alternative agriculture. For example, members of the Council on Agricultural Science and Technology (CAST) panel on organic farming were informed by the chairman that the purpose of the panel was not to evaluate the effectiveness, limitations and applicability of organic farming, but rather to show that it is thoroughly and entirely inappropriate (Busch and Lacy 1983).

Nevertheless, a recent surge in research by agroecologists has shown that traditional farming systems are often based on deep ecological rationales and in many cases exhibit a number of desirable features of socioeconomic stabililty, biological resilience and productivity (Egger 1981, Gliessman et al. 1981). Findings have also revealed that many organic farming systems conserve resources and are reasonably productive and profitable (USDA 1980).

Although modern high-input agricultural technology offers increased yields, in most cases it is not economically acceptable to traditional farmers in the developing world (Harwood 1979). On the other hand, an increasing number of so-called organic farmers in the U.S. and Europe reject such technologies on the basis that their prolonged use causes environmental degradation (Lockeretz et al. 1981). If the needs and concerns of these farmers are to be met, cropping systems that depend on low inputs of energy and resources and that are sustainable in the long term must be developed. Such systems should exhibit effective soil nutrient-restoring features as well as built-in biological pest control mechanisms.

These new agroecosystems should be based on existing systems from which basic principles can be extracted. One major frustration encountered by researchers in the United States is that the conditions necessary to develop alternative, sustainable agroecosystems simply are not present (Edens et. al. 1982). It is extremely difficult to find sites that have minimal energy inputs, lack continuous disturbances, and that are being managed with a primary goal of increasing the interactions among as many biological components as possible, rather than simply increasing yield.

In the developing countries, however, traditional farming systems constitute such settings to a great degree. These complex systems are well adapted to the local conditions, which have allowed peasants to meet their subsistence needs for centuries. If western agriculturalists are to learn about these farming systems, they must do so soon, before this wealth of practical knowledge is irretrievably lost.

"Alternative agriculture" is defined here as any approach to farming that attempts to provide sustained yields through the use of ecologically sound management technologies. Strategies rely on ecological concepts, such that management results in optimum recyling of nutrients and organic matter, closed energy flows, balanced pest populations and enhanced multiple use of the landscape.

The scientific discipline that approaches the study of agriculture from an ecological perspective is herein termed "agroecology" or "agricultural ecology" and is defined as a theoretical framework aimed at understanding agricultural processes in the broadest manner. The agroecological approach regards farm systems as the fundamental units of study, and in these systems, mineral cycles, energy transformations, biological processes and socioeconomic relationships are investigated and analyzed as a whole. Thus, agroecological research is concerned not with maximizing production of a particular commodity, but rather with optimizing the agroecosystem as a whole. This approach shifts the emphasis in agricultural research away from disciplinary and commodity concerns and toward complex interactions among and between people, crops, soil and livestock.

The purpose of this book is to provide a simple synthesis of the research on novel agroecosystems and technologies and an analysis of ecologically based farms, for the purpose of establishing the scientific basis of alternative agriculture. This book is an urgent call to all students of agriculture to seriously look at and learn from the numerous success stories of farmers who have managed their farms in a healthy and sustainable manner.

In the following chapters we will explore how this more "holistic" perspective can serve as the basis for developing more environmentally sound agricultural production systems. The book is divided into five parts. Part One describes the historical and theoretical framework of agricultural ecology. Part Two deals with the ecological considerations necessary in designing sustainable agroecosystems and suggests a methodology for evaluating farming systems for the purpose of designing technologies adapted to the needs and resources of alternative farmers. Part Three describes the ecological features of various traditional and organic farming systems throughout the world, showing that there are many living models to learn from, both for researchers and farmers. The purposeful blending of traditional and modern knowledge is the starting point in the development of a sustainable agriculture. Part Four shows the ecological basis for managing insect pests, pathogens and weeds. Part Five depicts the necessary conditions for the adoption of a sustainable agriculture worldwide.

The idea for this book emerged at the Ecological Farming Conference held by the California Steering Committee on Sustainable Agriculture at La Honda, California, in January 1983. The book undoubtedly has been influenced heavily by my experiences with traditional farming in Latin America and organic farming in California. My ecological and political views on agriculture emerged as I grew up in the midst of the controversies of the underdeveloped world, and later matured as I studied and worked in the contrasting culture of the United States.

Miguel A. Altieri

Acknowledgments

I am indebted to literally hundreds of people: To Juan Gasto, Jerry Doll, E. V. Komarek, W. H. Whitcomb, and W. J. Lewis, who gave me opportunities to explore new ideas and thus realize my potential; to many small farmers in Chile, Colombia, Mexico, Indonesia, and other areas, who have shown me centuries-tested methods of ecological agriculture; and to many students and colleagues at Berkeley and other universities in the United States and around the world, who have stimulated and challenged me to the utmost to think about new avenues in agricultural research and development.

I have also received essential encouragement and constructive criticism from my friends at the Division of Biological Control, University of California at Berkeley, with whom I share the same mission, if not neccessarily the same opinion. The University of California Appropriate Technology Program was exemplary in providing funds to produce the first printing. Steve Gliessman provided much valuable published and unpublished material that I used to complement various chapters. My colleagues and friends Richard Norgaard, John Farrell, Matt Liebman and Susanna Hecht each contributed one important chapter, thus enriching the concepts of agroecology explored in this book. I am also indebted to the Jessie Smith Noyes Foundation for supporting my agroecology research and training program at Berkeley and also in Chile for the last four years.

To Kat Anderson I owe thanks for constructive criticism, editorial assistance and intellectual and emotional support. Linda Schmidt drew many of the graphs, and her assistance in assembling the final drafts for printings of both editions was crucial. Catharine Way provided invaluable and effective assistance as copy editor for the second edition. Joanne Fox also lent her invaluable technical expertise and skill at all stages and printed

the final copy.
I owe appreciation to the following authors and publishers for permission to reproduce figures, tables and condensed versions of written materials:

—K. F. Wiersum, Wageningen Agricultural University, for material from the publication "Viewpoints on Agroforestry."

—C. Gregory Knight, Pennsylvania State University, for sections of the chapter on the nkomanjila and nkule systems from his book "Ecology and Change."

—Stephen R. Gliessman, University of California at Santa Cruz, for a condensed version of his paper "The ecological basis for the application of traditional agricultural technology in the management of tropical agroecosystems."

—Manuel C. Palada, Rodale Research Center, for the use of a summarized version of his paper "Association of interseeded legume cover crops and annual row crops in year—round cropping systems."

—Oekan Abdoelah, Padjadjaran Unversity, for material appearing in a paper he co-authored titled "Traditional Agroforestry in West Java."

—Academic Press, Inc. (London), for the use of Table 8.1 from the book "The Biology of Agricultural Systems," by C. R. W. Spedding.

—The American Society of Agronomy, for the use of Tables 1 and 2 from the chapter by Andrews and Kassam from ASA Publication No. 27.

—AVI Publishing Company, for material from Chapter 7 of the book by Thorne and Thorne, "Soil, Water and Crop Production."

—Elsevier Scientific Publishing Company, for use of material from the article by J. F. Parr et al., "Organic farming in the U.S.: principles and perspectives."

—Springer—Verlag, New York, for mention of material from the article by D. H. Mueller et al., "Conservation tillage: best management practice for non—point runoff."

—Maruja Salas of Minka magazine, for the drawing used on front cover of the book.

M.A.A.

PART ONE

The Theoretical Basis of Agricultural Ecology

1

The Evolution of Agroecological Thought

Susanna B. Hecht

> So in natural science, it is the composite thing, the thing as a whole which primarily concerns us, not just the materials of it, which are not found apart from the thing itself.
>
> Aristotle

HISTORICAL BACKGROUND

The contemporary use of the term agroecology dates from the 1970s, but the science and the practice of agroecology are as old as the origins of agriculture. As researchers explore indigenous agricultures, which are modified relics of earlier agronomic forms, it is increasingly apparent that many locally developed agricultural systems routinely incorporate mechanisms to accommodate crops to the variability of the natural environment and to protect them from predation and competition. These mechanisms make use of regionally available renewable inputs and ecological and structural features of the agricultural field, fallows and surrounding vegetation.

Agriculture in these situations involves managing resources other than the "target" crop. These production systems were developed to even out environmental and economic risk and maintain the productive base of agriculture over time. While such agroecosystems can include infrastructure like terraces, trenches and irrigation works, the decentralized, locally developed agronomic knowledge is central to the continuing performance of these production systems.

Why this agricultural heritage has been relatively unimportant in the formal agronomic sciences reflects biases that some contemporary researchers are trying to overcome. Three historical processes have done much to obscure and denigrate the agronomic knowledge that was developed by local peoples and non-western societies: (1) the destruction of the means of encoding, regulating and transmitting agricultural practices; (2) the dramatic transformation of many non-western indigenous societies and the production systems on which they were based as a result of demographic collapse, slaving and colonial and market processes; and (3) the rise of positivist science. As a result, there have been few opportunities for the insights developed in a more holistic agriculture to "filter up" into the formal scientific community. This difficulty is further compounded by unrecognized biases of agronomic researchers related to social factors such as class, ethnicity, culture and gender.

Historically, agricultural management included rich symbolic and ritual systems that often served to regulate land use practices, and to encode the agrarian knowledge of non-literate peoples (Ellen 1982, Conklin 1972). The existence of agrarian cults and ritual is documented for many societies, including those of Western Europe. Indeed, these cults were an essential focus of the Catholic Inquisition. Medieval social historians such as Ginzburg (1983) have shown how rural ceremonies were branded as witchcraft, and how such activities became the focus for intense persecution. Not surprisingly, as the post-Inquisition Spanish and Portuguese explorers set sail and European conquest spread over the globe for "God, gold and glory," part of their larger project included evangelical activities that often altered the symbolic and ritual bases of agriculture in non-western societies. These modifications transformed and often interfered with the generational and lateral transfer of local agronomic knowledge. This process, along with diseases, slaving and the frequent restructuring of the agricultural base of rural communities for colonial and market purposes, often contributed to the destruction or abandonment of the "hard" technologies such as irrigation systems, and especially to the impoverishment of "soft" technologies (cultivar types, cropping mixes, techniques of biological control and soil management) of the local agricultures, which are far more dependent on cultural forms of transmission.

The historical literature documents how the diseases carried by explorers affected native populations. Especially in the New World, rapid and unimaginably devastating population collapses occurred. As much as 90 percent of the population of some areas died in less than 100 years (Denevan 1976). With them died cultures and knowledge systems. The grisly effects of epidemics characterized the earlier phases of contact, but other activities, especially slaving for New World plantations, were also to have drastic impacts on population and thus on agricultural knowledge until well into the 19th century.

Initially, local populations were the focus of slave raids, but these groups were often able to escape from bondage. The disease problems of the New World Indians also made them a less than ideal labor force. African populations, on the other hand, were accustomed to tropical conditions, and relatively resistant to "European" diseases. They could thus satisfy the burgeoning manpower needs for sugar and cotton plantations. Over two centuries more than 20 million slaves were transported from Africa to various slave plantations in the New World (Wolf 1982).

Slaving is directed at the best labor force (young adult men and women) and it resulted in the loss of this important labor force for local agriculture and the abandonment of agricultural works as people sought to avoid slavery by moving to areas distant from slavers. The disruption of knowledge systems through the export of labor, the erosion of the cultural basis of local agricultures and the mortality associated with warfare stimulated by slaving raids was later compounded by the integration of these residual systems into mercantile and colonial networks.

The European contact with much of the non-western world was not benign, and often involved the transformation of productive systems to satisfy the needs of local bureaucratic centers, mining or resource enclaves, and international commerce. This was achieved through direct coercion in some cases, reorienting or manipulating economies through the collusion of existing local elites and headmen in others, and through exchange. These processes fundamentally change the basis of the agricultural economy. With the emergence of cash cropping and increased pressure on particular export items, rural land use strategies that had evolved over millenia to reduce agricultural risk and maintain the resource base were destabilized. Many studies have documented these effects (Watts 1983, Wolf 1982, Palmer and Parsons 1977, Wasserstrom 1982, Browkenshaw et al. 1979, Geertz 1962).

Finally, even when chroniclers and explorers made positive mention of native land use practices, it was difficult to translate these observations into a coherent, non-folkloric and socially acceptable form. The rise of the positivist method in science and the movement of western thought to atomistic and mechanistic perspectives (see Chapter 2) associated with the 18th century enlightment dramatically altered the discourse about the natural world (Merchant 1980).

This transition in epistemologies shifted the view of nature from that of an organic, living entity to one of a machine. Increasingly, this approach emphasized a language of science, a way of talking about the natural world that essentially dismissed other forms of scientific knowledge as superstitions. Indeed, from the time of Condorcet and Comte, the rise of science was equated with the triumph of reason over superstition. This position, coupled with an often derogatory view of the abilities of rural peoples generally, and colonized populations in particular, further obscured the richness of many rural knowledge systems whose content was expressed in discursive and symbolic form. By misunderstanding the ecological context, the spatial and cultivar complexity of non-formalized agricultures was frequently reviled as disorder.

Given this history, one might ask how agroecology managed to re-emerge at all. The "rediscovery" of agroecology is an unusual example of the impact of pre-existing technologies on the sciences, where critically important advances in the understanding of nature resulted from the decision of scientists to study what farmers had already learned how to do (Kuhn 1979). Kuhn points out that in many cases, scientists succeeded in "merely validating and explaining, not in improving, techniques developed earlier."

How the idea of agroecology re-emerged also requires the analysis of the influence of a number of intellectual currents that had relatively little to do with formal agronomy. The study of indigenous classification systems, rural development theory, nutrient cycling and succession have little direct relation to crop science, soil science, plant pathology and pest management as they are normally practiced. How disciplines as diverse as anthropology, economics and ecology are reflected in the intellectual pedigree of agroecology is outlined briefly in the next sections in this chapter, but the entire volume shows the influences on agroecological approaches in far more detail.

WHAT IS AGROECOLOGY?

The term agroecology has come to mean many things. Loosely defined, agroecology often incorporates ideas about a more environmentally and socially sensitive approach to agriculture, one that focuses not only on production, but also on the ecological sustainability of the production system. This might be called the "normative" or "prescriptive" use of the term agroecology, because it implies a number of features about society and production that go well beyond the limits of the agricultural field. At its most narrow, agroecology refers to the study of purely ecological phenomena within the crop field, such as predator/prey relations, or crop/weed competition.

The Ecological View

At the heart of agroecology is the idea that a crop field is an ecosystem in which ecological processes found in other vegetation formations—such as nutrient cycling, predator/prey interactions, competition, commensalism and successional changes-also occur. Agroecology focuses on ecological relations in the field, and its purpose is to illuminate the form, dynamics and function of these relations. Implicit in some agroecological work is the idea that by understanding these processes and relations, agroecosystems can be manipulated to produce better, with fewer negative environmental or social impacts, more sustainably, and with fewer external inputs. As a result, a number of researchers in the agricultural sciences and related fields have begun to view the agricultural field as a particular kind of ecosystem—an agroecosystem—and to formalize the analysis of the ensemble of processes and interactions in cropping systems. The underlying analytic framework owes much to systems theory and the theoretical and practical attempts at integrating the numerous factors that affect agriculture (Spedding 1975, Conway 1981, Gliessman 1982, Conway 1985, Chambers 1983, Ellen 1982, Altieri 1983, Lowrance et al. 1984).

The Social Perspective

Agroecosystems have various degrees of resiliency and stability, but these are not strictly determined by biotic or environmental factors. Social factors such as a collapse in market prices or changes in land tenure can disrupt agricultural systems as decisively as drought, pest outbreak or soil nutrient decline. On the other hand, decisions that allocate energy and material inputs can enhance the resiliency and recuperation of damaged ecosystems. Although human manipulations of ecosystems for agricultural production have often dramatically altered the structure, diversity, patterns of energy and nutrient flux and mechanisms of population regulation within agricultural fields, these processes still operate, and can be explored experimentally. The magnitude of the differences in ecological function between a natural and an agricultural ecosystem depends tremendously on the intensity and frequency of the natural and human perturbations that impinge on an ecosystem. The results of the interplay between endogenous biological and environmental features of the agricultural field, and exogenous social and economic factors, generate the particular agroecosystem structure. For this reason, a broader perspective is often needed to explain an observed production system.

An agricultural system differs in several fundamental ways from a "natural" ecological system in its structure and function. Agroecosystems are semi-domesticated ecosystems that fall on a gradient between ecosystems that have experienced minimal human impact, and those under maximum human control, like cities. Odum (1984) describes four major characteristics of agroecosystems:

1. Agroecosystems include auxiliary sources of energy like human, animal and fuel energy to enhance productivity of particular organisms.
2. Diversity can be greatly reduced compared with many natural ecosystems.
3. The dominant animals and plants are under artificial rather than natural selection.
4. The controls on the systems are largely external rather than internal via subsystem feedback.

Odum's model is primarily based on modernized agriculture, such as that found in the United States. There are, however, many kinds of agricultural systems, particularly in the tropics, that do not fit well with this definition. Particularly suspect are questions of diversity and the nature of selection in complex agricultures when a number of semi-domesticates and wild plants and animals figure into the production system. For example, Conklin (1956) described agroecosystems in the Philippines that included over 600 cultivated and managed plants. While this agriculture was not as diverse as some tropical forests, it was certainly more diverse than many other local ecosystems.

Agricultural systems are complex interactions between external and internal social, biological and environmental processes. These can be understood spatially at the level of the agricultural field, but often include a temporal dimension as well. The degree of external vs. internal control can reflect intensity of management over time, which can be far more variable than Odum suggests. In swidden systems, for example, external controls tend to drop off in the later fallow periods. Odum's model of agroecosystems is an interesting point of departure for understanding agriculture from an ecological systems perspective, but cannot capture the diversity and complexity of many agroecosystems that evolved in non-western societies, particularly in the humid tropics. Moreover, the model's lack of attention to the social determinants of agriculture results in a model with limited explanatory power.

Agricultural systems are human artifacts, and the determinants of agriculture do not stop at the boundaries of the field. Agricultural strategies respond not only to environmental, biotic and cultivar constraints, but also reflect human subsistence strategies and economic conditions (Ellen 1982). Factors like labor availability, access and conditions of credit, subsidies, perceived risk, price information, kinship obligations, family size and access to other forms of livelihood are often critical to understanding the logic of a farming system. Particularly when analyzing situations of small farmers outside the U.S. and Europe, simple yield maximization in monocultural systems becomes less useful for understanding farmer behavior and agronomic choices (Scott 1978 and 1986, Bartlett 1984, Chambers 1984).

The Agroecology Challenge

Conventional agricultural scientists have been concerned primarily with the effect of soil, animal or vegetation management practices upon the productivity of a given crop, using a perspective that emphasized a target problem such as soil nutrients or pest outbreaks. This means of addressing agricultural systems has been determined in part by the limited dialogue across disciplinary lines, by the structure of scientific investigation, which tends to atomize research questions, and by an agricultural commodity focus. There is no question that agricultural research based on this approach has been successful in increasing yields in favored situations.

Increasingly, however, scientists are recognizing that such a narrow approach could limit agricultural options for rural peoples, and that the "target approach" often carries with it unintended secondary consequences that have often been ecologically damaging and had high social costs. Agroecology research does concentrate on target issues in the agricultural field, but within a wider context that includes ecological and social variables. In many cases, premises about the purposes of an agricultural system may be at variance with the purely productionist or yield focus of some agricultural scientists.

Agroecology can best be described as an approach that integrates the ideas and methods of several subfields, rather than as a specific discipline. Agroecology can be a normative challenge to existing ways of approaching agricultural issues in several disciplines. It has roots in the agricultural sciences, in the environmental movement, in ecology (particularly in the explosion of research on tropical ecosystems), in the analysis of indigenous agroecosystems and in rural development studies. Each of these areas of inquiry has quite different aims and methodologies, yet taken together, they have all been legitimate and important influences on agroecological thought.

INFLUENCES ON AGROECOLOGICAL THOUGHT

Agricultural Sciences

As Altieri (1983) has pointed out, credit for much of the initial development of agricultural ecology in the formal sciences belongs to Klages (1928), who suggested that consideration be given to the physiological and agronomic factors influencing the distribution and adaptation of specific crop species to understand the complex relationships between a crop plant and its environment. Later Klages (1942) broadened his definition to include the historical, technological and socioeconomic factors that determined what crops could be produced in a given region and in what amount. Papadakis (1938) stressed that the culture of crops should be based on the crops' response to environment. Agricultural ecology was elaborated further in the 1960s by Tischler (1965) and integrated into the agricultural curriculum in which courses were oriented to developing the basis for an ecological point of view on crop adaptation. Agronomy and crop ecology are increasingly converging, but the networks between agronomy and the other sciences (including social sciences) necessary for agroecological work are just coming into being.

The works of Azzi (1956), Wilsie (1962), Tischler (1965), Chang (1968) and Loucks (1977) represent the gradual shift toward an ecosystem approach to agriculture. Azzi (1956) in particular emphasized that while meteorology, soil science and entomology are distinct disciplines, their study in relation to the potential responses of crop plants converges in an agroecological science that should illuminate the relationships between crop plants and their environment. Wilsie analyzed the principles of crop adaptation and distribution in relation to habitat factors, and made an attempt to formalize the body of relationships implicit in crop systems. Chang (1968) further pursued the lines suggested by Wilsie, but focused to a greater degree on the ecophysiological aspects.

Since the early 1970s, there has been an enormous expansion of agronomic literature with an agroecological perspective, including works such as Dalton 1975, Netting 1974, van Dyne 1969, Spedding 1975, Loomis et al. 1971, Cox and Atkins 1979, Richards, P. 1984, Vandermeer 1981, Edens and Koenig 1980, Edens and Haynes 1982, Altieri and Letourneau 1982, Gliessman et al. 1981, Conway 1985, Hart 1979, Lowrance et al. 1984 and Bayliss-Smith 1982.

By the late 1970s and early 1980s an increasingly large social component appeared in the agricultural literature, largely as a result of rural development studies and critiques of U.S.

agricultural development structures (Buttel 1986). The social contextualization coupled with agronomic analysis has generated complex evaluations of agriculture, particularly in relation to regional development (Altieri and Anderson 1986, Brush 1977, Richards, P. 1984 and 1986, Kurin 1983, Bartlett 1984, Hecht 1985, Blaikie 1984).

Pest managers, particularly entomologists, have made important contributions to the development of an ecological perspective in plant protection. The theory and practice of biological pest control is based on ecological principles (Huffaker and Messenger 1976). Ecological pest management focuses primarily on approaches that contrast the structure and function of agricultural systems with those of relatively undisturbed systems or more complex agricultural systems (Southwood and Way 1970, Price and Waldbauer 1975, Levins and Wilson 1979, Risch 1981 and Risch et al. 1983). Browning (1975) has argued that pest management approaches should emphasize the development of agroecosystems that emulate later successional stages as much as possible, since these types of systems are often more stable than systems of simple monocultural structure.

Methodological Approaches. A great deal of agroecological analysis in agricultural sciences is currently under way throughout the world. At this juncture four main methodological approaches are routinely used:

1. Analytic description. Many studies are under way that carefully measure and describe agricultural systems and measure particular properties such as plant diversity, biomass accumulation, nutrient retention and yields. For example, the International Center on Agroforestry (ICRAF) has been developing an international data base on the various types of agroforestry systems, and is correlating them with a variety of environmental parameters to develop systematic regional and crop models (Nair 1984, Huxley 1983). This kind of information is valuable for expanding our understanding of the types of existing systems, which components are normally found together and in what environmental context. It is the necessary first step. Representative studies along these lines are numerous and include Ewel et al. 1984, Alcorn 1984, Marten 1986, Turner and Brush 1986, Deneven et al. 1984 and Posey 1985.

2. Comparative analysis. Comparative research usually involves comparing a monoculture or other cropping system with a more complex traditional agroecosystem. Comparative studies like this involve analysis of productivities of particular crops, pest dynamics or nutrient status as they correlate with factors like crop field diversity, weed frequencies, insect populations and nutrient cycling patterns. Several such studies have been carried

out in Latin America, Africa and Asia (Glover and Beer 1986, Uhl and Murphy 1981, Irvine 1987, Marten 1986 and Woodmansee 1984). Such projects use standard scientific methodologies to illuminate the dynamics of particular local mixed cropping systems compared with monocultures. These data are often useful, but the heterogeneity of local systems may obscure how they are functioning.

3. <u>Experimental comparison.</u> To clarify the dynamics and reduce the number of variables many researchers develop a simplified version of an indigenous system in which the variables can be more closely controlled. For example, yields of an intercrop of corn, beans and squash can be compared with pure stands of each of these crops.

4. <u>Normative agricultural systems.</u> These are often constructed with particular theoretical models in mind. A natural ecosystem can be mimicked, or an indigenous agriculture system could be painstakingly reconstituted. This approach is being experimentally evaluated by several researchers in Costa Rica. They are developing cropping systems that emulate the successional sequences by using cultivars that are similar botanically or morphologically to plants in various successional sequences (Hart 1979, Ewel 1986).

While agronomy most certainly has been the mother discipline of agroecology, it was strongly influenced by the emergence of environmentalism and the expansion of ecological studies. Environmentalism was necessary to provide the philosophical framework on which the value of the alternative technologies and the normative project of agroecology could rest. Ecological studies were critical to expand the paradigms through which agricultural questions could be developed, and the technical skills for analyzing them.

Environmentalism

<u>Importance of the movement.</u> A major intellectual contributor to agroecology has been the environmental movement of the 1960s and 1970s. As environmental issues translated into agroecology they infused parts of the agroecology discourse with a critical stance toward production-oriented agronomy, and increased sensitivity to a broad range of resource issues.

The 1960s version of the environmental movement arose initially out of concern about pollution issues. These were analyzed as a function of both technology failures and population pressures. The Malthusian perspective gained particular force in the mid-1960s with works like Paul Ehrlich's <u>The Population Bomb</u> (1966) and Garrett Hardin's <u>Tragedy of the Commons</u> (1968). These authors linked environmental degradation and resource depletion primarily to population increases. This point of view was expanded technically by the publication of the Club of Rome's The Limits to Growth, which used computer simulations of global trends in population, resource use and pollution to generate scenarios for the future, which were generally disastrous. This position has been critiqued from methodological and epistemological perspectives (Simon and Kahn 1985).

While The Limits to Growth developed a generalized model of the "environmental crisis," two later seminal volumes had particular relevance to agroecological thought because they outlined visions of an alternative society. These were the Blueprint for Survival (The Ecologist, 1972) and Schumacher's Small is Beautiful (1973). The works incorporated ideas about social organization, economic structure and cultural values into comprehensive, more or less utopian visions. Blueprint for Survival argued for decentralization, smallness of scale and an emphasis on human activities that would involve minimal ecological disruption and maximum conservation of energy and materials. The passwords were self-sufficiency and sustainability. Schumacher's book emphasized a radical evaluation of economic rationality ("Buddhist economics"), a decentralized model of human society ("two million villages") and appropriate technology. Of particular significance in Small is Beautiful was the extension of these ideas to the Third World.

Agricultural questions. The environmental issues as they pertained to agriculture were clearly signaled by Carson's Silent Spring (1964), which raised questions about the secondary impacts of toxic substances, especially insecticides, in the environment. Part of the response to these problems was the development of pest management approaches to crop protection that were in theory and practice based entirely on ecological principles (Huffaker and Messenger 1976). Toxicity of agrochemicals was only one of the environmental questions, since energy resource use was also becoming an increasingly important topic. The energy costs of particular production systems required evaluation, particularly early in the 1970s when oil prices skyrocketed. Pimentel's 1973 classic study showed that in American agriculture, each kilocalorie of corn was "purchased" at enormous energetic cost of external energy. U. S. production systems were subsequently compared with several other forms of agriculture that were less productive per unit area (in terms of kilocalories per hectare) but much more efficient in terms of return per unit of energy expenditure. The high yields of modern agriculture are purchased at the price of numerous inputs including non-renewable inputs like fossil fuel and phosphorous.

In the Third World, these inputs are often imported, and strain the international balance of payments and debt situation of many developing countries. Further, because food crops do not receive most of these inputs, production gains may not translate into a better food supply (Crouch and de Janvry 1980, Graham 1984, Dewey 1981). Finally, the social consequences of this model have complex and often extremely negative impacts on local populations, particularly those with limited access to land and credit. These issues are discussed in more detail later in this chapter, and elsewhere in this volume.

The toxicity and resource issues in agriculture dovetailed with the larger questions of technology transfer in Third World contexts. The Careless Technology (edited by Milton and Farver in 1968) was one of the early major attempts to document the effects of development projects and the transfer of temperate-zone technologies on the ecologies and societies of developing countries. Increasingly, researchers from several fields began to comment on the poor "fit" between First World land use approaches and Third World realities. Janzen's 1973 article on tropical agroecosystems was the first widely read evaluation of why tropical agricultural systems might function differently from those of the temperate zones. This chapter and that of Levins (1973) was a challenge to agricultural researchers to rethink the ecology of tropical agriculture.

At the same time, the larger philosophical issues raised by the environmental movement resonated with the reevaluation of the goals of agricultural development in the U.S. and the Third World, and the technological basis on which these would be carried out. In the developed world these ideas had only moderate impact on the structure of agriculture, because the reliability and availability of agrochemical and energetic inputs to agriculture resulted in minor transformations in the patterns of resource use in agriculture. In situations where farmers and nations were constrained by resources, where regressive distributional structures prevailed and where temperate zone approaches were often inappropriate for local environmental conditions, the agroecology approach seemed particularly relevant.

The integration of agronomy and environmentalism dovetailed in agroecology, but the intellectual foundations for such an academic mix were still relatively weak. A clearer theoretical and technical approach was needed, particularly with respect to tropical systems. The developments in ecological theory were to have particular relevance to the evolution of agroecological thought.

Ecology

Ecologists have been singularly important in the evolution of agroecological thought for several reasons. First, the conceptual framework of agroecology and its language are essentially ecological. Second, agricultural systems are themselves interesting research ensembles where researchers have much greater ability to control, test and manipulate the components of the system compared with natural ecosystems. These can provide test conditions for a wide array of ecological hypotheses, and indeed have already contributed substantially to the body of ecological knowledge (Levins 1973, Risch et al. 1983, Altieri et al. 1983, Uhl et al. 1987). Third, the explosion of research on tropical ecosystems has drawn attention to the ecological impacts of expanding monocultural systems in zones characterized by extraordinary diversity and complexity (Janzen 1973, Uhl 1983 or Uhl and Jordan 1984, Hecht 1985). Fourth, a number of ecologists have begun to turn their attention to the ecological dynamics of traditional agricultural systems (Gliessman 1982a,b, Altieri and Farrell 1984, Anderson et al. 1987, Marten 1986, Richards, P. 1984 and 1986).

Three areas have been particularly critical in the development of agroecological analyses: nutrient cycling, pest/plant interactions and succession. Pest/plant interactions and successional issues are dealt with in more detail throughout the volume so this section will focus primarily on nutrient cycling.

In the early 1960s, nutrient cycling analysis of tropical ecosystems became a focus of interest in the tropics and as a vital ecosystem process, given the general poverty of many tropical soils. Several studies such as Odum's Puerto Rico Study (1967), Nye and Greenland's 1961 research and later the series of articles and monographs derived from the work at San Carlos, Venezuela; Catie, Costa Rica; and other Asian and African sites have been seminal in illuminating the mechanisms of nutrient cycling in both native forests and cleared areas (Jordan 1985, Uhl and Jordan 1984, Buschbacker et al. 1987, Uhl et al. 1987).

The ecological findings of this nutrient cycling research that had the greatest impact on agricultural analysis were:

- o The relationship between diversity and interspecies nutrient strategies.
- o The importance of structural features for enhancing nutrient capture both above and below ground.

o   The dynamics of physiological mechanisms for nutrient retension.
o   The importance of associative relations of higher plants with microorganisms such as mycorrhiza and symbiotic nitrogen fixers.
o   The importance of biomass as the site of nutrient storage.

These findings suggested that ecological models of tropical agriculture would include a diversity of species (or at least cultivars) to take advantage of the variability of nutrient uptake, both in terms of different nutrients and in capturing nutrients from different depths of the solum. The information generated from ecological studies of nutrient cycling also suggested the use of plants that readily form symbiotic associations, such as legumes, and the more widespread use of perennials in the production system, as a means of nutrient pumping from different depths of the soil and to increase the total ecosystem nutrient storage capacity. Not surprisingly, many of these principles were already in practice in numerous agricultural systems developed by local populations in the tropics.

In most ecological literature the comparison of natural ecosystems with agroecosystems has been based on agroecosystems developed by ecologists after some observation of local ecosystems, rather than truly locally evolved ones. Moreover, research questions focused on parameters like seed diversity, biomass accumulation and nutrient storage in succession. This research has provided us with an understanding of some of the dynamics of agricultural systems as biological entities, but how management (except that carried out by relatively inexperienced graduate students) influences these processes remains an enormously unexplored area. (For a salient exception in this regard see Uhl et al. 1987).

The limitations of the purely ecological approach are being increasingly overcome as researchers begin to examine peasant and indigenous systems in multidisciplinary teams and from a more holistic perspective (Anderson and Anderson 1983, Hecht et al. 1987, Anderson et al. 1987, Marten 1986, Denevan et al. 1984). These efforts attempt to put agriculture in a social context; they use indigenous local models (and indigenous explanations for why they do particular activities) for developing hypotheses that can then be tested using agronomic and scientific methods. This is a burgeoning research area with major theoretical and applied implications, and a major inspiration to agroecological theory and practice.

Indigenous Production Systems

Another major influence on agroecological thought has come from the research efforts of anthropologists and geographers concerned with describing and analyzing the agricultural practices and logic of indigenous and peasant peoples. These studies have typically been concerned with resource use and management of the entire subsistence base, not just the agricultural plot, and have focused on how this subsistence base is explained by local peoples and how social and economic change affect production systems. The scientific analysis of local knowledge has been an important force in reevaluating the assumptions of colonial and agricultural development models. The pioneer work of this type was that of Audrey Richards (1939) on the <u>citamene</u> swidden practices of the African Bemba. The <u>citamene</u> system involves using tree litter as compost in agriculture of the scrub woodlands of central Africa. This study, with its emphasis on the outcomes of agricultural technologies and ecological explanations of native people, stood in stark contrast to the derogatory perceptions of native agriculture, which viewed local practices as messy and inferior.

Another major contribution on indigenous cultivation was the seminal work by Conklin (1956), which laid the groundwork for reevaluating shifting cultivation based on ethnographic and agronomic data on the Hanunoo of the Philippines. This work pointed out the ecological complexity of shifting cultivation patterns as well as the diversity of types of shifting cultivation, and the importance of multi-cropping, rotation cropping and agroforestry systems in the overall shifting cultivation production framework. It is among the earliest and most widely known studies of the structure and complexity of shifting cultivation, and incorporates many ecological insights.

Of particular importance was Conklin's emphasis on native ecological knowledge, and the importance of tapping this rich source of scientific understanding. He emphasized, however, that access to this information would require ethnographic as well as scientific skills.

Researchers such as Richards, P. 1984, Bremen and deWit 1983, Watts 1983, Posey 1984, Denevan et al. 1984, Hecht and Posey 1987, Browkenshaw et al. 1979, and Conklin 1956, among many others, have explored indigenous systems of production and categories of knowledge about environmental conditions and agricultural practices. This body of research focuses on the native view of production systems, and analyzes them with the methods of western science. All of these authors have emphasized that social organization and the social relations of production should be considered as closely as the environment and cultivars. This emphasis on the social dimensions of production is an important basis for understanding the production logic of agricultural systems.

Another important result of much of the work on indigenous production systems is the idea that different notions of efficiency and rationality are required to understand indigenous and peasant systems. For example, efficiency of output per unit of labor investment, rather than a simple output per area ratio, is basic to the production logic of many Third World cultivators. Practices that focus on risk aversion may not be high yielding in the short run, but may be preferable to highly productive and risky land options. Labor availability, particularly at seasonal pulses like harvesting, may also influence the types of agricultural systems that are favored.

This kind of research has been influential in developing the counter arguments to those who attributed the failure of agricultural technology transfers to ignorance and indolence. This approach, with its emphasis on the human factors in agricultural systems, also focused greater attention on the strategies of peasants of different class strata, and increasingly on the role of women in agriculture and resource management (Deere 1982, Beneria 1984, Moock 1986).

Ethnoagricultural analysis has done much to expand the conceptual and practical tool kit of agroecology. The focus on "emic" frameworks (based on a given culture's explanation) has suggested relationships that "etic" frameworks (that is, external frameworks, usually referring to western models of explanation) would not easily capture, but that can often be tested with the methods of western science. Moreover, this research has expanded the conception of what can usefully be considered agriculture, as many groups are engaged in forest ecosystem manipulation through the management of succession and actual reforestation (Posey 1985, Anderson et al. 1987, Alcorn 1984). Moreover, locally developed agriculture incorporates numbers of cultivars whose germplasm is essential to "developed" breeding programs like those of manioc and beans, and also includes numbers of plants with the potential for more widespread use in difficult environments. Finally, such work valorizes the scientific achievements of hundreds of years of plant breeding and agronomic work by local peoples.

The study of indigenous agricultural systems has provided much of the raw material for the development of hypotheses and alternative production systems for agroecology. Native agriculture is now increasingly studied by multidisciplinary teams to document practices, and classification categories have been developed to analyze the biological processes within agricultural systems, and to evaluate aspects of the social forces that influence agriculture. The study of indigenous systems has been seminal in the development of agroecological thought.

Development Studies

The study of rural Third World development has also made a major contribution to the evolution of agroecological thought. Rural analysis has helped clarify the logic of local production strategies in communities undergoing intense transformation, as rural areas increasingly are integrated into larger regional, national and global economies. Rural development studies have documented the relations between socioeconomic factors and the structure and social organization of agriculture. Several themes in development research have been particularly important in agroecology, including the impacts of externally induced technology and cropping change, the effects of market expansion, the implications of changes in social relations and the transformation in tenure structures and access to customary resources. All of these processes are deeply interwoven. How they affect regional agroecosystems is a result of complex historical and political processes.

Research on the Green Revolution was important in the evolution of agroecological thought because studies of the impact of this technology were instrumental in illuminating the types of biases that predominated in agricultural and development thinking. This research also resulted in the first really multidisciplinary analysis of ecological, social and economic tenure issues and technical change in agriculture by a broad spectrum of analysts. The extraordinary acceleration in peasant social stratification associated with the Green Revolution indicated immediately that this was not a scale-neutral technology, but one that could dramatically transform the basis of rural life for large numbers of people.

As noted in Perelman 1977, the major beneficiaries of such technologies were urban consumers. The Green Revolution strategy evolved when the problems of poverty and hunger were viewed primarily as problems of production. This diagnosis implied several strategies that focused on the areas where production gains could be realized rapidly: better quality soils and irrigated lands among farmers with substantial assets. In terms of raising output, it succeeded; at bottom it was part of a policy of betting consciously on the strong (Chambers and Ghildyal 1985, Pearse 1980). It is now generally recognized that aggregate increases in food production alone will not overcome <u>rural</u> starvation and poverty, although it may reduce some urban food costs (Sen 1981, Watts 1983).

The Green Revolution had consequences in rural areas that often served to marginalize much of the rural population. First, it focused its benefits on the already resource-rich farmers, accelerating the differentiation between them and other rural inhabitants, so rural inequality often increased. Second, it undermined many forms of access to land and resources such as share cropping, labor tenancies and access to water supplies and grazing lands. This reduced the diversity of subsistence strategies available to rural households and thus increased their dependence on the agricultural plot. With the narrowing of the genetic basis of agriculture, risk increased because crops became more vulnerable to pest or disease outbreaks and the vagaries of climate. In irrigated rice, the secondary pollution generated by the increased use of pesticides and herbicides often undermined the important source of local protein: fish.

The analysis of the Green Revolution from several disciplines constituted the first holistic analysis of agricultural/rural development strategies. It was the first widely publicized evaluation that incorporated ecological, technological and social critiques. This kind of approach and analysis has been the prototype for several subsequent agroecological studies, and the progenitor of farming systems research.

It is now well recognized that Green Revolution technologies can be applied in limited areas, and there have been calls by several rural development analysts to redirect agricultural research toward resource-poor farmers. Worldwide there are more than one billion such farmers with very limited assets, income and production flows who work in an agricultural context of extreme marginality. Agricultural approaches emphasizing technological packages have generally required resources to which most of the world's farmers have no access (Table 1.1).

Many rural development analysts had recognized the limitations of large-scale and Green Revolution-oriented approaches to rural development, but these agricultural models have overwhelmingly dominated agricultural development projects in much of the Third World. While research station results looked extremely promising, the weak replicability of these results in the field has caused serious difficulty in many projects. The transfer-of-technology approach tended to accelerate differentiation, exacerbating many difficult political situations, or the technologies were partially adopted, and in many cases they were not adopted at all (Scott 1978 and 1986).

Several explanations accounted for the poor transfer of technology, including the idea that farmers were ignorant and

Table 1.1 The contrast in physical and socio-economic conditions of resource-rich vs. resource-poor farmers (modified from Chambers and Ghildyal, 1985)

	Research Station	Resource Rich Farmers (RRF)	Resource Poor Farmers (RPF)
Topography	flat or terraces	flat or terraces	undudated or sloped
Soils	deep, few constraints	deep, few constraints	shallow, infertile serious constraints
Nutrient deficiency	rare, remedial	occasional	quite common
Hazards (fire, land slides, etc.)	few	few controllable	common
Irrigation	often, full control	usually available reliable control	rare, unreliable
Size of unit	large, contiguous	large or medium contiguous	small, irregular often non-contiguous
Disease, Pests, Weeds	controlled with chemicals, labor	controlled w/chemicals labor	crops vulnerable to infestation
Access to fertilizers, Improved seed etc.	unlimited, reliable	high, reliable	low, unreliable
Seeds	high quality	high quality	own seed
Credit	unlimited	good access	poor access with seasonal shortages
Labor	no constraint	controlled by farmer; hired	family, constraining at seasonal peaks
Prices	irrelevant	lower for inputs, higher for outputs relative to RPF	higher for inputs lower for outputs relative to RRF
Priority for food production	—	low	high

needed to be taught how to farm. Another set of explanations focused on farm-level constraints such as lack of credit that limited the ability of farmers to adopt technologies. In the first case the farmer is viewed as basically at fault. In the second, infrastructural questions of various types are considered the culprit. Never was the technology itself criticized.

Many field researchers and development practitioners have been frustrated by these explanations, and have increasingly indicated that the technologies themselves require substantial reevaluation. They have argued that the farmer's decision to adopt a technology is the true test of its quality. This approach has often been called "the farmer first and last" or "farmer back to farmer," or "indigeneous agricultural revolution." As Rhodes and Booth (1982) put it, "The basic philosophy upon which the model is based is that agricultural research and development must begin and end with the farmer. Applied agricultural research cannot begin in isolation out on the research station or with a planning committee out of touch with farm conditions. In practice, this means obtaining information about and understanding the farmers' perception of the problem, and accepting farmers' evaluation of the solution." This approach calls for much broader farmer participation in the design and implementation of rural development programs (Chambers 1984, Richards, P. 1984, Gow and Van Sant 1983, Midgley 1986).

One consequence of this stance has been the recognition of the very extensive knowledge of farmers in entomology, botany, soils and agronomy, which can serve as the starting point for research. Here again, agroecology has been identified as a valuable analytical tool as well as a normative approach to research.

Agroecology fits well with the technological issues requiring more environmentally sensitive agriculture practices, and often finds congruencies in both environmental and participatory development in philosophical perspectives. The diversity of concerns and bodies of thought that have influenced the development of agroecology is broad indeed. However, this is the range of issues that impinges on agriculture. It is for this reason that we now see agroecologists with much richer training than is usual for students of agricultural sciences, and many more multidisciplinary teams dealing with these issues in the field. While it is a discipline in its infancy, and so far has raised more questions than solutions, it has widened the agricultural discourse.

2

The Epistemological Basis of Agroecology

Richard B. Norgaard

Epistemology, or the philosophy of knowledge, is concerned with how scientists think they know. Scientists rarely contemplate how they know unless a crisis occurs, such as their experiments repeatedly indicate something contrary to existing knowledge, their knowledge does not work as expected when applied in the real world or another way of knowing begins to challenge openly the existing way.

Western agricultural science is not in a crisis, but it has not always worked out as expected. Chemical pest controls have had numerous unforeseen secondary impacts. New varieties of crops requiring more fertilizer and water have resulted in unexpected impacts on soils and groundwater supplies. As a result, many agricultural scientists acknowledge a "mild crisis" because knowledge derived by modern science has not worked as expected when applied in the real world.

Agroecology contributes to this "mild crisis" with a challenge to the dominant western way of thinking. The agroecological way of knowing is quite different. Agroecologists are fascinated by agricultural systems that have evolved over centuries, in which people are actively involved. Agroecologists study how people interact in these systems and learn about important relationships through the farmers' explanations of why they farm as they do.

Most agricultural scientists, however, conduct controlled laboratory experiments or undertake novel tests in experimental plots where the vagaries of normal agricultural practices are carefully eliminated. Explanations derive from the accumulation of knowledge from past research and the information provided by the additional experiments. These general differences in approach are rooted in basic differences in ways of knowing.

CHARACTERISTICS OF WESTERN THOUGHT

Early western scientists set out to understand a static world as God had created it. They envisioned the acquisition of knowledge as a process in which individual minds investigate nature's parts and processes. In western tradition the mind has been thought of as an independent entity that perceives and interprets. Asking questions, thinking and acting neither influence the underlying principles that govern nature nor affect the mind itself. Like the mind, nature in the dominant world view also just exists. The world just is and the mind just perceives and interprets. In this way, people and the natural world have been juxtaposed from the beginning of western thought. The emphasis on the objectivity of knowledge stems from this static juxtaposition of mind and nature.

Western thought on knowledge has several other important characteristics. First, there has always been a strong emphasis on useful knowledge. Second, western science is always interested in universal phenomena. These two characteristics are complementary, for universal knowledge is more useful because it can be applied anywhere. Third, the world can be perceived as consisting of many atomistic parts that can be described and "known" independently of each other. Fourth, the parts are related in a systematic manner that can be known. Knowing, in this case, entails being able to predict the effect on the whole system of a change in one of the parts. Prediction requires that the system can be described in a manner that is manageable and logical, so that "if this, then that" statements can be made. For knowledge to be universal, neither the nature of the parts nor the nature of the relationships can change. The relative proportion of the parts and the relative strengths of the relationships, however, can change.

This western world view might be illustrated as in Figure 2.1. We observe nature, derive theories about universal characteristics of natural parts and relationships in nature, test theories against nature, design technologies and social organization based on theories and thereby modify nature. But we only modify the proportions of the parts and relative strengths of the relationships. The universal nature of the parts and relations remains unchanged. With nature unchanging, knowledge about nature can accumulate over time. The idea that science is cumulative, that we are coming closer and closer to knowing all there is to know about the parts and relations in nature, is deeply embedded in western thought. Once in a while, scientists lament that there may be an infinite amount to know, but few question that we do not continue to know more about the unchanging principles of nature. This belief in the accumulation of scientific knowledge complements the western belief in progress.

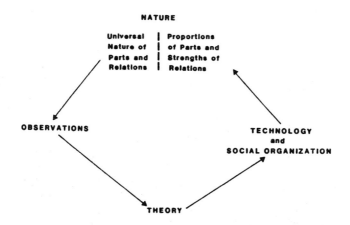

Figure 2.1   The dominant, or mechanical, world view.

Our dominant world view is referred to as mechanical partly because Newton's mechanics was the most important breakthrough in this way of thinking, and partly because of analogies with machines. The parts and relationships of a machine do not change, but the relative strengths of the relationships do change when a machine changes speed, turns or stops.

THE AGROECOLOGICAL WORLD VIEW

Agroecologists use knowledge from western agricultural science to help them understand agroecosystems. But agroecologists know they are interpreting complex systems that have evolved with people as part of a unique process, rather than a machine with universal characteristics that operates apart from people. The laws of physics with respect to motion, the attractive forces and heat are universal. The ways in which molecules chemically react do not vary. But the numbers of ways in which the simple parts and relationships studied by the physicist and chemist can combine to form complex biological organisms, let alone ecological systems with human actors, is infinite. Agroecologists perceive how each organism has evolved within the context of a larger evolving system. Though species may travel between regions, the evolutionary path of the larger system remains unique to the area.

The most important difference between the agroecological world view and that of western science is that agroecologists perceive people as a part of evolving local systems. The nature of each biological system has evolved to reflect the nature of the people—their social organization, knowledge, technologies and values. People have selected characteristics of species for centuries. People have helped maintain desirable biological relationships. What species and varieties are selected and which relationships are assisted depends on people's values, what they know, how they are socially organized to interact with their environment and biological system, and the techniques available to them.

Similarly, people's natures reflect some of the characteristics of the physical environment and biological system. Different physical terrains and climates and their associated biological systems—alpine, tropical rainforest, savanna and desert—lead to different ways of knowing, select for different forms of social organization, support different technologies and encourage different values. People have evolved differently in different environments and biological systems.

And so human culture molds biological systems while biological systems mold culture. Each puts selective pressure on the other. People and their biological systems have coevolved. The ecosystem, in this view, includes the knowledge system, value system, social organization and technology of the people along with the biological system. This, of course, is a much bigger ecosystem than most ecologists are willing to ponder. It is illustrated in Figure 2.2.

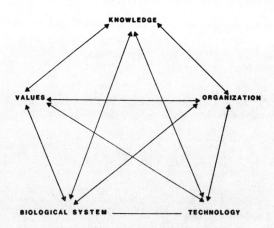

Fig. 2.2 The coevolution of knowledge, values, social organization, technology and the biological system.

Within this broad coevolutionary world view, how do agroecologists know? They do not know universal truths, for each agroecosystem they study has a different coevolutionary history. They know that the nature of the parts can only be understood in the context of the coevolution of the whole. Parts and relations change with time, relatively little time in the case of small species and the genetic diversity of larger species. These changes reflect the decisions of people conditioned by their values, beliefs, organization and technology. As a result, the concepts of objective knowledge and objectivity are moot.

Similarly, what is the meaning of progress in western knowledge, when people are trying to understand something that is changing, and how it changes is related to how people know? Though agroecologists have a more sophisticated view of the world, how the emerging "discipline" knows can only be juxtaposed against the epistemological beliefs of conventional western thought. Epistemological beliefs for an evolutionary world view are still in the making.

BLENDING WESTERN AND "TRADITIONAL" KNOWLEDGE

In the absence of a consensus about epistemological beliefs, agroecologists have resorted to pragmatism. Western knowledge is not rejected, for the mechanical world view has given us many insights, and conventional agricultural explanations help the agroecologist understand traditional systems as well. At the same time, agroecologists are receptive to the explanations of traditional peoples. Traditional knowledge may not survive western tests. Traditional knowledge may not generate testable hypotheses; when it does, the hypotheses may be refuted; and the knowlege—typically contained in myths and social expectations—may not even be internally consistent. But traditional knowledge has survived the test of time—the selective pressures of droughts, downpours, blights and pest invasions—and usually for more centuries than western knowledge has survived.

The evolutionary world view is slowly affecting how scientists think in fields beyond agroecology (Grene 1985; Quinne and Dunham 1983, Sober 1984). In the social sciences, especially anthropology, researchers have always been frustrated by the pretense of objectivity, by mechanical explanations of evolutionary processes and by the fitness of alternative world views. Agroecology shares its epistemological base with the anthropological subdiscipline of cultural ecology where evolution of culture is explained with reference to the environment and the evolution of the environment is explained with reference to culture (Durham 1978, Netting 1974, Steward 1977). Sociobiologists have also explored the epistemological base of people as a part of nature (Lorenz 1977, Lumsden and Wilson 1981). Geographers are independently describing traditional agricultural systems and critiquing agricultural modernization in agroecological frameworks (Richards 1984). Economists, philosophers, systems theorists and many others are exploring evolutionary approaches to understanding (Boulding 1979, Norgaard 1981 and 1984, Toulmin 1982, Bateson 1979). Agroecology may be part of a broader turning point in western thought (Berman 1981, Capra 1982, Merchant 1983).

The scope of agroecology has not yet evolved implicit boundaries through practice, nor has a definition of the field been adopted by a professional scientific body. The epistemological explorations of this chapter, however, suggest the following set of premises that explicitly define the field:

1. Biological and social systems, as systems, have agricultural potential.
2. This potential has been captured by traditional farmers through a process of trial, error, selection and cultural learning.
3. Social and biological systems have coevolved such that each depends upon feedback from the other. Knowledge, embodied in traditional cultures through cultural learning, stimulates and regulates the feedback from social to biological systems.
4. The nature of the potential of social and biological systems can best be understood, given the present state of formal social and biological knowledge, by studying how traditional farming cultures have captured the potential.
5. Formal social and biological knowledge, the knowledge and some of the inputs developed by conventional agricultural sciences, and experience with western agricultural technologies and institutions can be combined to improve both traditional and modern agroecosystems.
6. Agricultural development through agroecology will maintain more cultural and biological options for the future and have fewer detrimental cultural, biological and environmental effects than conventional agricultural science approaches alone.

These six premises together constitute a minimum set necessary to provide limits, define a world view, suggest the approach and define aspirations for agroecology. Each of the premises has been explicitly argued, though in different words, or implicitly assumed in the agroecological literature to date. Conventional western sciences have shorter definitions only because they share commonly accepted, unstated, epistemological roots and theories as to how science relates to development.

While agroecologists accept these premises, they do not necessarily accept all of them all the time. They have hesitated to delineate their field too precisely, to draw unnecessary distinctions where distinctions may only antagonize. Specifically, by putting less emphasis on cultural learning and social and biological coevolution, agroecology blends smoothly into an ecosystems approach, a social systems approach or an integrated approach to agriculture. Each of these approaches generates information useful to agroecology and, because they are already accepted, provides avenues to apply agroecological knowledge.

But these approaches differ in that they assume objective knowledge can be used rationally to design modern agricultural systems. The deemphasis of premises (2) and (3) reduces the barriers between agroecologists and systems-oriented agricultural scientists interested in alternatives such as agroforestry, polycropping, farm household decision-making or farming systems. Nevertheless, it is the cultural knowledge and coevolutionary premises that make agroecology unique, controversial and a productive contribution to our understanding of science and development.

Agroecology has a different epistemological basis than most of western science. To have different roots is to be radical in the true sense of the word. Conventional scientists strive to bring new technologies derived from modern science to traditional farmers so they might become "developed." Agroecologists strive to understand how traditional systems "developed" to enhance the science of ecology so that modern agriculture might be made more sustainable. Agroecologists, in short, are removing the "one way" signs along the road between science and development.

The epistemological differences between western science and agroecology could lead to an unhealthy conflict and cause a potentially fruitful new way of thinking to be "nipped in the bud." At the same time, the differences could lead to some exciting lines of questioning and highly productive research. If the scientific community can harbor two epistemological bases of thought, the next several decades could become innovative periods for both the agricultural sciences and for development policy.

NOTES

1. Associate Professor of Agriculture and Resource Economics, University of California, Berkeley.

3

The Agroecosystem: Determinants, Resources and Processes

The terms agroecosystem, farming system and agricultural system have been used to describe agricultural activities performed by groups of people. Food system is a broader term that includes agricultural production, allocation of resources and product processing and marketing within an agricultural region and/or country (Krantz 1974). Obviously an agroecosystem can be defined at any scale, but this book focuses primarily on agricultural systems within small geographical units. Thus, the emphasis is on interactions between people and food-producing resources within a farm or even a specific field. It is difficult to delineate the exact boundaries of an agroecosystem. Nevertheless, it should be kept in mind that agroecosystems are open systems receiving inputs from outside and producing outputs that can enter external systems (Figure 3.1).

CLASSIFICATION OF AGROECOSYSTEMS

Each region has a unique set of agroecosystems that result from local variations in climate, soil, economic relations, social structure and history (Table 3.1). Thus, a survey of the agroecosystems of a region is bound to yield both commercial and subsistence agricultures, using high or low levels of technology depending on the availability of land, capital and labor. Some technologies in the more modern systems aim at land saving (relying on biochemical inputs), while others emphasize labor saving (mechanical inputs). Traditional, resource-poor farmers usually adopt more intensive systems, emphasizing optimal use and recycling of scarce resources.

Although each farm is different, many show a family likeness and can thus be grouped together as a type of agriculture, or agroecosystem. An area with similar types of agroecosystems can

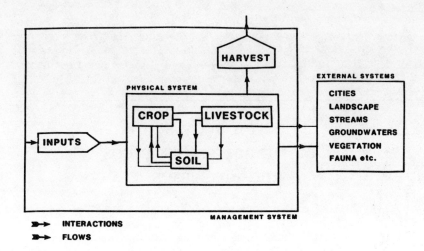

Figure 3.1 The general structure of an agricultural system and its relationship with external systems (after Briggs and Courtney 1985).

Table 3.1 Agroecosystem determinants that determine the type of agriculture in each region

Type of Determinants	Factors
Physical	Radiation Temperature Rainfall, water supply (moisture stress) Soil conditions Slope Land availability
Biological	Insect pests and natural enemies Weed communities Plant and animal diseases Soil biota Background natural vegetation Photosynthetic efficiency Cropping patterns Crop rotation
Socio-economic	Population density Social organization Economic (prices, markets, capital and credit availability) Technical assistance Cultivation implements Degree of commercialization Labor availability
Cultural	Traditional knowledge Beliefs Ideology Gender issues Historical events

then be termed an agricultural region. Whittlesay (1936) recognized five criteria to classify agroecosystems in a region: (1) The crop and livestock association; (2) the methods used to grow the crops and produce the stock; (3) the intensity of use of labor, capital and organization, and the resulting output of product; (4) the disposal of the products for consumption (whether used for subsistence on the farm or sold for cash or other goods); and (5) the ensemble of structures used to house and facilitate farming operations.

Based on these criteria, in tropical environments it is possible to recognize seven main types of agricultural systems (Grigg 1974, Norman, M. 1979):

1. Shifting cultivation systems
2. Semi-permanent rainfed cultivation systems
3. Permanent rainfed cultivation systems
4. Arable irrigation systems
5. Perennial crop systems
6. Grazing systems
7. Systems with regulated ley farming (alternating arable cropping and sown pasture).

Clearly these systems are always changing, forced by population shifts, resource availability, environmental degradation, economic growth or stagnation, political change, etc. These changes can be explained by farmers' responses to variations in the physical environment, prices of inputs and products, technological innovation and population growth. For example, Table 3.2 illustrates some of the factors influencing the change from shifting cultivation systems to more intensive permanent systems of agriculture in Africa (Protheroe 1972).

THE RESOURCES OF AN AGROECOSYSTEM

Norman (1979) grouped the mix of resources commonly found in an agroecosystem into four categories:

Natural Resources

Natural resources are the given elements of land, water, climate and natural vegetation that are exploited by the farmer for agricultural production. The most important elements are the area of the farm, including its topography, the degree of fragmentation of the holding, its location with respect to markets; soil depth, chemical status and physical attributes; availability of

Table 3.2 Factors influencing agricultural intensification in African regions where shifting cultivation is practiced (Protheroe 1972)

FACTOR	PROCESS
POPULATION	LOW DENSITY → *Increasing numbers* → HIGH DENSITY Possible surplus rural population Seasonal/Permanent migration → Urban drift
SYSTEM	SHIFTING CULTIVATION → ROTATIONAL CULTIVATION FALLOW → SEMI-PERMANENT/PERMANENT CULTIVATION → *Increasing length of cultivation period* → *Decreasing length of fallow period* → *Manuring and fertilizing*
CROPS	SUBSISTENCE FOOD CROPS → *Decreasing importance* → CASH (FOOD AND EXPORT) CROPS → *Increasing importance* →
TENURE	COMMUNAL RIGHTS TO LAND → *Communal rights decreasing* → INDIVIDUAL RIGHTS TO LAND (individual usufructary rights) *individual rights increasing* Land allocation by need → Land transfer by pledge, rent, lease and sale Fragmented/dispersed holdings → Consolidated holdings No permanent demarcation of holdings → Permanent demarcation of holdings
SETTLEMENT	IMPERMANENT/MIGRATORY SMALL VILLAGES/DISPERSED → *Increasing permanence and nucleation* → PERMANENT/FIXED NUCLEATED AND DISPERSED
EXCHANGE	NONEXISTENT/LOCAL → *Increasing involvement at local, regional, national and international levels* → MARKETS

surface water and ground water; average rainfall, evaporation, solar radiation and temperature (and its seasonal and annual variability); and natural vegetation, which may be an important source of food, animal feed, construction materials or medicines for humans, and an influence on soil productivity in shifting cultivation systems.

Human Resources

The human resources consist of the people who live and work within the farm and exploit its resources for agricultural production, based on their traditional or economic incentives. The factors affecting these resources include (a) the number of people the farm has to support in relation to the work force and its productivity, which governs the surplus available for sale, barter or cultural obligations; (b) the capacity for work, as influenced by nutrition and health; (c) the inclination to work, as influenced by economic status and cultural attitudes toward leisure; and (d) the flexibility of the work force to adapt to seasonal variations in work demand, that is, the availability of hired labor and the degree of cooperation among farms.

Capital Resources

Capital resources are the goods and services created, purchased or borrowed by the people associated with the farm to facilitate their exploitation of natural resources for agricultural production. Capital resources can be grouped into four main categories: (a) permanent resources, such as lasting modifications to the land or water resources for the purpose of agricultural production; (b) semipermanent resources, or those that depreciate and have to be replaced periodically, like barns, fences, draft animals, implements; (c) operational resources, or consumable items used in the daily operations of the farm, like fertilizer, herbicides, manure and seeds; and (d) potential resources, or those the farmer does not own but that may be commanded and that will eventually have to be repaid, like credit and assistance from relatives and friends.

Production Resources

Production resources include the agricultural output of the farm such as crops and livestock. These become capital resources when sold, and residues (crops, manure) are nutrient inputs reinvested in the system.

ECOLOGICAL PROCESSES IN THE AGROECOSYSTEM

Every farmer must manipulate the physical and biological resources of the farm for production. Depending on the degree of technological modification, these activities affect four major ecological processes: energetic, hydrological, biogeochemical and biotic regulation processes. Each can be evaluated in terms of inputs, outputs, storages and transformations.

Energetic Processes

Energy enters an agroecosystem as sunlight and undergoes numerous physical transformations. Biological energy is transferred into plants by photosynthesis (primary production) and from one organism to another through the food web (consumption). Although sunlight is the only major source of energy input in most natural ecosystems, human and animal labor, mechanized energy inputs (such as plowing with a tractor) and the energy content of introduced chemicals (manures, fertilizers and pesticides) are also significant. Human energy shapes the structure of the agroecosystem, thereby shaping energy flow through decisions about primary production and the proportion of that production that is channeled to products for human use (Marten 1986).

The various inputs into an agricultural system—solar radiation, human labor, the work of machines, fertilizers and herbicides—can all be converted into energy values. Similarly the outputs of the system—the various vegetable and animal products—can also be expressed in energy terms. As the cost and availability of fossil fuel energy are questioned, inputs and outputs have been quantified for different kinds of agricultures to compare their intensity, yields and labor productivity and the levels of welfare they provide. In his comparative analysis of seven types of agricultural systems, Bayliss-Smith (1982) found that the overall efficiency of energy use (energy ratio) diminishes as dependence on fossil fuels increases. Thus, in fully industrialized agriculture, the net gain of energy from agriculture is small because so much is expended in its production (Figure 3.2).

Biogeochemical Processes

The major biogeochemical inputs into an agroecosystem are the nutrients released from the soil, fixation of atmospheric nitrogen by legumes, nonsymbiotic nitrogen fixing (particularly important in rice growing), nutrients in rainfall and run-on water, fertilizer and nutrients in purchased human food, stock feed or animal manure.

The important outputs include nutrients in crops and livestock consumed on or exported from the farm. Other outputs or losses are associated with leaching beyond the root zone, dentrification and volatilization of nitrogen, losses of nitrogen and sulfur to the atmosphere when vegetation is burned, nutrients lost in soil erosion caused by runoff or wind and nutrients in human or livestock excreta that are lost from the farm.

There is also biogeochemical storage, including the fertilizer stored and manure accumulated, together with the nutrients in the soil root zone, the standing crop, vegetation and livestock.

In the course of production and consumption, mineral nutrients move cyclically through an agroecosystem. The cycles of some of the most important nutrients (nitrogen, phosphorous and potassium) are well understood in many natural and agricultural ecosystems (Todd et al. 1984). During production, elements are transferred from the soil into the plants and animals, and vice versa. Whenever carbon chains are broken apart through a variety of biological processes, nutrients are returned to the soil where they can sustain plant production (Marten 1986, Briggs and Courtney 1985).

Farmers move nutrients in and out of the agroecosystem when they bring in chemical or organic fertilizers (manure or compost) or remove the harvest or any other plant materials from the field. In modern agroecosystems, lost nutrients are replaced with purchased fertilizers. Low-income farmers who cannot afford commercial fertilizers sustain soil fertility by collecting nutrient

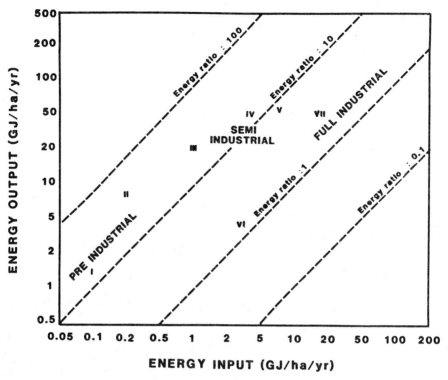

Figure 3.2 Inputs, outputs and energy ratios of seven agricultural systems. I: Traditional morning farming in New Guinea (shifting cultivation, home gardens), II: British pre-industrial farming system (grain/sheep system), III: Ongoing Java agricultural system (taro gardens, coconut woodland, fishing), IV: South India pre-Green Revolution (sugar cane, rice, finger millet, bullock grazing), V: South India post-Green Revolution (sugar cane, rice, finger millet, bullock grazing), VI: Russian collective farm (potatoes, grain, grazing), VII: Modern British agriculture (grains, ley and permanent grass) (Bayliss-Smith 1982).

materials from outside the crop fields, such as manure collected from pastures or enclosures in which animals are kept at night. This organic material is supplemented with leaves and other plant materials from nearby forests. In areas of Central America farmers spread as much as 40 metric tons of litter per hectare each year over intensively cropped vegetable fields (Wilken 1977). Waste plant materials are composted with household wastes and manure from livestock.

Another strategy to exploit the ability of the cropping system to reuse its own stored nutrients. In interplanted agroecosystems the low disturbance and closed canopies promote nutrient conservation and cycling (Harwood 1979). For example, in an agroforestry system, minerals lost by annuals are rapidly taken up by perennial crops. In addition, the nutrient-robbing propensity of some crops is counteracted by the addition of organic matter from other crops. Soil nitrogen can be increased by incorporating legumes in the mixture, and phosphorous assimilation can be enhanced somewhat in crops with mycorrhizal associations. Increased diversity in cropping systems is usually associated with larger root area, which increases nutrient capture.

Hydrological Processes

Water is a fundamental part of all agricultural systems. In addition to its physiological role, water affects inputs of nutrients to and losses from the system through leaching and erosion. Water enters an agroecosystem as precipitation, run-on and irrigation water; it is lost through evaporation, runoff and drainage beyond the effective root zone of plants. Water consumed by the people and livestock on the farm may be important (such as in pastoral systems) but is usually small in magnitude.

Water is stored in the soil, where it is used directly by crops and vegetation, in groundwater that may be drawn up for use by people, livestock or crops, and in constructed storages such as farm ponds.

In general terms, the water balance within a particular agroecosystem can be expressed as:

$$S = R + L_i - E_t - P - L_o + S_o$$

where S is the soil moisture content at the time under consideration, R is effective rainfall (rainfall minus interception), L_i is the lateral flow of water into the soil, E_t is evapotranspiration, P is deep percolation, L_o is the lateral outflow (runoff) and S_o is the original soil moisture content (Norman, M. 1979; Briggs and Courtney 1985).

All these factors are affected by soil and vegetation conditions, and thus by agricultural practices. Agricultural drainage and tillage, for example, speed up losses by deep percolation; crop removal increases the amount of rainfall reaching the soil and reduces evapotranspiration; changes in soil structure due to tillage residue management, crop rotation or use of manure affect rates of percolation, evapotranspiration and lateral flow. One of the main controls of the soil moisture budget is exerted by crop cover, for it influences both inputs to and losses from soil moisture. For example, weeding reduces water losses from evapotranspiration and increases soil moisture contents.

In rainfed agriculture, it is important to know that when R is greater than E_t the root zone is fully charged, and defines the effective crop growing season. During this period runoff and drainage can also occur, influencing the level of leaching of soluble nutrients, rate of soil erosion, etc. Within the range of $R = E_t/2$ to $R = E_t/10$, continued crop growth and maturation depend largely on the available soil water reserve or on irrigation (Norman, M. 1979).

In most rainfed tropical areas the agricultural potential of the area depends on the length of the rainy season and the distribution of rainfall during this period. Satisfactory crop climates are those in which rainfall exceeds actual evapotranspiration for at least 130 days, the length of an average growing cycle for most annual crops. The number of consecutive wet months is another important environmental criterion. The potential for sequential cropping (under rainfed conditions) is limited if there are less than five consecutive wet months (Beets 1982).

Rainfall is a major determinant of the type of crops adopted in the local cropping system. In Africa, where annual precipitation is more than 600 mm, cropping systems are generally based on maize. In tropical Asia, where precipitation is more than 1,500 mm/year with at least 200 mm/month rainfall for three consecutive months, cropping systems are generally based on rice. Since rice needs more water than other crops, and because it is the only crop that tolerates flooding, only rice is grown at the peak of the rains. A combination of upland crops can be planted at the beginning or end of the rains to use residual moisture and higher light intensities during the dry season (Figure 3.3). Mixed cropping systems such as maize and groundnuts, for example, often best use the end of the rainy season (System II in Figure 3.3).

Another possibility is to combine a double and relay cropping system in which transplanted rice is established as early as possible (System III in Figure 3.3). The rice is followed by cowpeas raised using minimum tillage techniques, and cucurbits are relay-planted later (Beets 1982).

Biotic Regulation Processes

Controlling succession (plant invasion and competition) and protecting against insect pests and diseases are major problems in maintaining production continuity in agroecosystems. Farmers have used several approaches universally. These are no action, preventive action (use of resistant crop varieties, manipulation of planting dates, row spacing, modifying access of pests to plants) or suppressive action (chemical pesticides, biological control,

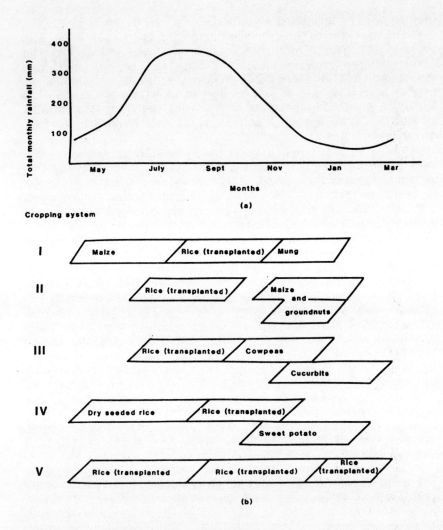

Figure 3.3 Five possible cropping systems that fit a rainfall pattern in Southeast Asia (Beets 1982).

cultural techniques). Ecological strategies of pest management generally employ a combination of all three approaches, aiming at making the field less attractive to pests, making the environment unsuitable to pests but favorable to natural enemies, interfering with the movement of pests from crop to crop or attracting pests away from crops. All these approaches will be discussed in

Chapters 13, 14 and 15 as they pertain to insect, weed and plant disease management in agroecosystems.

THE STABILITY OF AGROECOSYSTEMS

Under conventional agriculture, humans have simplified the structure of the environment over vast areas, replacing nature's diversity with a small number of cultivated plants and domesticated animals. This process of simplification reaches an extreme form in a monoculture. The objective of this simplification is to increase the proportion of solar energy fixed by the plant communities that is directly available to humans. The net result is an artificial ecosystem that requires constant human intervention. Commercial seed-bed preparation and mechanized planting replace natural methods of seed dispersal; chemical pesticides replace natural controls on populations of weeds, insects and pathogens; and genetic manipulation replaces natural processes of plant evolution and selection. Even decomposition is altered since plant growth is harvested and soil fertility maintained, not through nutrient recycling, but with fertilizers. Although modern agroecosystems have proven capable of supporting a growing population, there is considerable evidence that the ecological equilibrium in such artificial systems is very fragile.

Why Modern Systems are Unstable

The explanation for this potential instability must be sought in terms of changes imposed by people. These changes have removed crop ecosystems from the natural ecosystem to the extent that the two have become strikingly different in structure and function (Table 3.3). Natural ecosystems reinvest a major proportion of their productivity to maintain the physical and biological structure needed to sustain soil fertility and biotic stability. The export of food and harvest limits such reinvestment in agroecosystems, making them highly dependent on external inputs to achieve nutrient cycling and population regulation (Cox and Atkins 1979).

Some researchers believe that biotic diversity and structural complexity provide a natural, mature ecosystem with a measure of stability in a fluctuating environment (Murdoch 1975). For example, severe stresses in the external physical environment, such as a change in moisture, temperature or light, are less likely to harm the entire system because in a diverse biota, numerous alternatives exist for the transfer of energy and nutrients. Hence, the system can adjust and continue to function after stress with

Table 3.3 Structural and functional differences between natural ecosystems and agroecosystems (modified from Odum 1969)

Characteristics	Agroecosystem	Natural ecosystem
Net productivity	High	Medium
Trophic chains	Simple, linear	Complex
Species diversity	Low	High
Genetic diversity	Low	High
Mineral cycles	Open	Closed
Stability (resilience)	Low	High
Entropy	High	Low
Human Control	Definite	Not needed
Temporal permanence	Short	Long
Habitat heterogeneity	Simple	Complex
Phenology	Synchronized	Seasonal
Maturity	Immature, early successional	Mature, climax

little if any detectable disruption. Similarly, internal biotic controls (such as predator/prey relationships) prevent destructive oscillations in pest populations, further promoting the overall stability of the natural ecosystem.

The modern agricultural strategy can be viewed as a reversal of the successional sequence of nature. These modern ecosystems, despite their high yield to humankind, carry with them the disadvantages of all immature ecosystems.

In particular, modern agroecosystems lack the ability to cycle nutrients, conserve soil and regulate pest populations. System functioning thus depends on continued human intervention. Even the crops selected for cultivation frequently cannot reproduce themselves without the assistance of humans, through sowing, and are incapable of competing against weed species without constant control. However, there is great variability in the degree of diversity, stability, human control, energy efficiency and productivity among the different types of agroecosystems (Figure 3.4).

Artificial Control in Modern Agroecosystems

To maintain normal levels of productivity in both the short term and the long term, modern agroecosystems require considerably more environmental control than organic or traditional

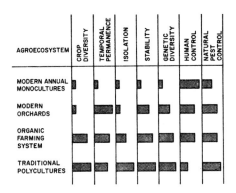

Figure 3.4 Ecological patterns of contrasting agroecosystems.

agricultural systems (Figure 3.5). The modern systems require large amounts of imported energy to accomplish the work usually done by ecological processes in less disturbed systems. Thus, although less productive on a per-crop basis than modern monocultures, traditional polycultures are generally more stable and more energy efficient (Cox and Atkins 1979). In all agroecosystems the cycles of land, air, water and wastes have become open, but it has happened to a larger degree in industrialized commercial monocultures than in diversified small-scale farming systems dependent on human/animal power and local resources.

These farming systems differ not only in their levels of productivity per area or per unit of labor or input, but also in more fundamental properties. It is apparent that, while new technology has greatly increased short-term productivity, it has also lowered the sustainability, equity, stability and productivity of the agricultural system (Figure 3.6) (Conway 1985). Those indicators are defined here as follows:

Sustainability refers to the ability of an agroecosystem to maintain production through time, in the face of long-term ecological constraints and socioeconomic pressures.

Equity is a measure of how evenly the products of the agroecosystem are distributed among the local producers and consumers (Conway 1985). However, equity is much more than simply a matter of an adequate income, good nutrition or a satisfactory amount of leisure (Bayliss-Smith 1982). To some, equity is reached when an agroecosystem meets reasonable demands

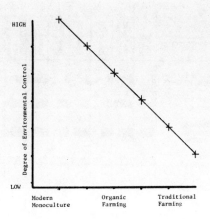

Figure 3.5 Degree of environmental control necessary for the maintenance of normal levels of productivity in three types of farming systems.

for food without increases in the social cost of production. To others, equity is reached when the distribution of opportunities or incomes within producing communities improves (Douglass 1984).

<u>Stability</u> is the constancy of production under a given set of environmental, economic and management conditions (Conway 1985). Some ecological pressures, like weather, are rigid constraints in the sense that the farmer virtually cannot modify them. In other cases the farmer can improve the biological stability of the system by choosing more suitable crops, or developing methods of cultivation that improve yields. The land can be irrigated, mulched, manured or rotated, or crops can be grown in mixtures to improve the resilience of the system. The farmer can supplement family labor with either animals or machines, or by employing other people's labor. Thus, the exact response depends on social factors as well as the environment. For this reason, the concept of stability must be expanded to embrace socioeconomic and management considerations. In this regard, Harwood (1979) defines three other sources of stability:

<u>Management stability</u> is derived from choosing the set of technologies best adapted to the farmers' needs and resources. Initially, industrial technology usually increases yield, as less and less land is left fallow, and soil, water and biotic limitations are bypasssed. But there is always an element of instability associated

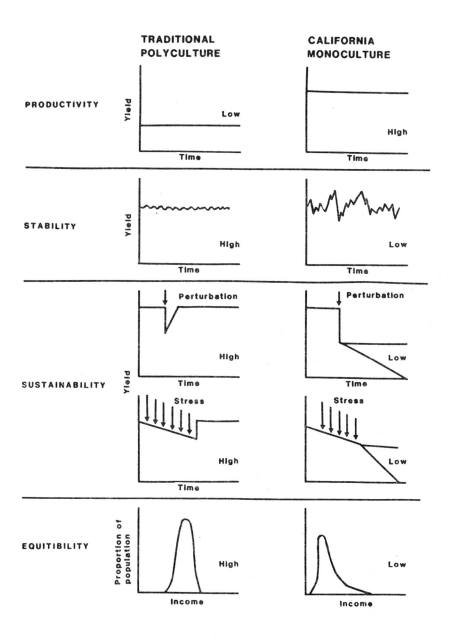

Figure 3.6 The system properties of agroecosystems and indices of performance (modified after Conway 1985)

with the new technologies. The farmers are keenly aware of this, and their resistance to change often has an ecological basis.

   Economic stability is associated with the farmer's ability to predict market prices of inputs and products, and to sustain farm income. Depending on the sophistication of this knowledge, the farmer will make tradeoffs between production and stability. To study the dynamics of economic stability in agricultural systems, data must be obtained on total production, yields of important commodities, cash flow, off-farm income, net income and the fraction of total production the farmer sells or trades.

   Cultural stability depends on the maintenance of the sociocultural organization and context that has nurtured the agroecosystem through generations. Rural development cannot be achieved when isolated from the social context, and it must be anchored to local traditions.

   Productivity is a quantitative measure of the rate and amount of production per unit of land or input. In ecological terms, production refers to the amount of yield or end product, and productivity is the process for achieving that end product. In evaluating small farm production, it is sometimes forgotten that most farmers place a higher value on reducing risk than on maximizing production. Small farmers usually are more interested in optimizing productivity of scarce farm resources than in increasing land or labor productivity. Also, farmers choose a particular production technology based on decisions made for the entire farming system, not only for a particular crop (Harwood 1979). Yield per unit area can be one indicator of the rate and constancy of production, but it can also be expressed in other ways, such as per unit of labor input, per unit of cash investment or as energy efficiency ratios. When patterns of production are analyzed using energy ratios, it becomes clear that traditional systems are exceedingly more efficient than modern agroecosystems in their use of energy (Pimentel and Pimentel 1979). A commercial agricultural system typically exhibits input/output ratios of three/one, whereas traditional farming systems exhibit ratios of one/10-15.

   The overall vulnerability of simplified modern agroecosystems is well illustrated by the epidemic of southern corn leaf blight that devastated the corn crop in the United States in 1970, and the destruction of millions of tons of wheat in the midwestern states in 1953 and 1954 by race 15B of Puccinia graminis f. sp. tritici (Baker and Cook 1974). The potato late-blight epidemic and subsequent famine in Ireland in the mid-19th century is a strong reminder that growing vast acreages of a highly simplified community is not a dependable means of food production. An alarming picture emerges from a report prepared by the National

Research Council of the National Academy of Sciences on the extent to which many staple crops have become genetically uniform and vulnerable to epidemics (Adams et al. 1971). This trend toward uniformity is apparent in the post-Green Revolution tendency for farmers to plant a single high-yielding variety in place of several different traditional varieties.

The intensification of agriculture is a crucial test of the resiliency of nature. We do not know how much longer humans can keep increasing the magnitude of the subsidy without depleting natural resources and causing further environmental degradation. Before we discover this critical point through unfortunate experience, we should endeavor to design agroecosystems that compare in stability and productivity with natural ecosystems (Cox and Atkins 1979). This is the driving force of agroecology.

PART TWO

The Design of Alternative Agricultural Systems and Technologies

4

Generating Sustainable Technologies

Over the past century, abundant resources, cheap energy, technological innovation and cultural factors have fueled agricultural growth in the industrialized countries. Most agricultural development projects have aimed at increasing production of agricultural commodities and linkages to markets (Perelman 1977, Conway 1985). This emphasis on increasing agricultural output was transferred to developing countries, without considering ecological and socioeconomic conditions, and it was justified by viewing the problems of rural poverty and hunger largely as problems of production. Consequently, techniques of agricultural development have not accounted for the needs and potentials of local peasants.

CONSEQUENCES OF INAPPROPRIATE TECHNOLOGY

Examples of the environmental consequences of dramatic technological changes abound in less-developed countries. One example is the substitution of tractor for buffalo power in Sri Lanka (Senanayake 1984). At first sight this substitution in the villages of Sri Lanka seemed to involve a simple tradeoff between more timely planting and labor saving on one hand, and the provision of milk and manure on the other. But buffaloes create buffalo wallows, and these in turn provide a surprising number of benefits. In the dry season they are a refuge for fish, which then move back to the ricefields in the rainy season. Some fish are caught and eaten by the farmers and by the landless, providing valuable protein; other fish eat the larvae of mosquitoes that carry malaria. The thickets surrounding the wallows harbor snakes that eat rats that eat rice, and lizards that eat the crabs that make destructive holes in the ricebuds. The wallows are also used by the villagers to prepare coconut fronds for thatching. If the wallows go, so do these benefits.

The adverse consequences may not stop there. If pesticides are used to kill the rats, crabs or mosquito larvae, pollution or pesticide resistance can become a problem. Similarly, if tiles are substituted for thatch, forest destruction may be hastened since firewood is required to fire the tiles (Conway 1986).

Another clear example of inappropriate technology is the Green Revolution, which attempted to solve crop production problems in the Third World through the development of high-yielding cereal varieties requiring massive inputs of pesticides, fertilizers, irrigation and machinery (Perelman 1977). Contrary to expectations, no significantly new technological packages capable of increasing yields could be offered to the majority of peasants (de Janvry 1981). The new packages failed to take into account the features of subsistence agriculture—ability to bear risk, labor constraints, symbiotic crop mixtures, diet requirements—that determine the management criteria and levels of resource use by local farmers. In the majority of cases, new varieties could not surpass local varieties when managed with traditional practices (Perelman 1977).

The areas where the new 'miracle cereals' were widely adopted were haunted by disease epidemics. Plant breeders soon learned that planting an entire region with genetically similar varieties could lead to disastrous attacks by either insect pests or diseases (Adams et al. 1971). Peasants soon abandoned the new varieties because of their high production costs (de Janvry 1981). For example, most small farmers could not afford the tubewell needed for irrigation, an essential component of the new technology (Perelman 1977). Thus, it seems that only a small proportion of farmers benefited from the Green Revolution.

ALTERNATIVE PRODUCTION SYSTEMS

In general terms, crop production can be increased by expanding the area planted to crops, by raising the yield per unit area of individual crops (usually by increasing input use) or by growing more crops per year in time and space.

Whatever strategy is used, crop yield (Y) is influenced by management (M), environment (E) and the genotype (G) of the crop (Beets 1982, Zandstra et al. 1981):

$$Y = f(M, E, G)$$

Management includes the cropping arrangement in time and space and cultural techniques. Environment is comprised of soil and climate variables modifiable through management. The crop

Table 4.1 Strategies to augment the yield of individual crops (after Beets 1982)

	Changing of genotype	Changing of crop environment through management	Requirements and results
Group I	Choice of already available crop species	Planting at the right time in the season Optimum plant populations spacing and configuration (including multiple cropping) Changing the microclimate Changing the competition with other living organisms	Requires little capital investment; can result in significant yield increases
Group II	Introduction of new crops	Timely and correct soil preparation Changing the soil conditions	Some capital may be required for import of seeds and implements
Group III	Selection and breeding of high yielding varieties of crop species already available	Use of fertilizers Introduction of irrigation Use of pesticides	Demands capital and know-know, but will give dramatic improvements when done well

genotype is inherent to the crop variety chosen and its adaptive range. Using this concept, three main strategies to augment the yield of individual crops can be recognized (Table 4.1). The applicability of each strategy will depend mainly on the availability of technical skills and capital relations (cost of inputs and prices for crops).

Although during the last decades greater emphasis has been placed on increasing yields per unit area through the use of labor-saving technologies (mechanization) and land-saving technologies (fertilizer, pesticides), recently agricultural scientists have become aware that it is important, not only to increase food production, but to do so with the most efficient use of energy and non-renewable resources (Wittwer 1975). Some promising approaches to agricultural technology, although valuable, have been based only on a single crop production process and have not considered the whole ecosystem. Table 4.2 provides a list of recommended and/or experimental practices for energy-efficient production.

For the most part, the more integrated approaches are directed toward enhancing photosynthetic efficiency through (a) improving plant architecture, using C_4 plants or varieties with a high leaf-area index, adopting efficient planting patterns, and hormonal stimulation of net photosynthesis; (b) improving soil management through minimum tillage, living legume mulches, cover cropping, manures, enhancement of biological N_2 fixation and use of mycorrhizae; (c) managing water more efficiently through drip irrigation, mulching and windbreaks; and (d) managing pests in an ecologically sound manner. These technologies propose minor changes in one or two components of the system, leaving the structure of the monoculture unchallenged, but without these minor changes, realistic progress cannot be made in the development of sustainable agroecosystems.

However, if the management boundaries are expanded beyond the direct object of control (i.e., a pest problem, soil nutrient deficiency, weed infestation) a whole new set of management and design options emerges (Edens and Koening 1981). Of special relevance are those manipulations that can simultaneously affect several components of the system. For example, growers who adopt novel agronomic systems (multiple cropping or agroforestry systems) can achieve several crop management objectives simultaneously and sometimes require little if any fertilizer or pesticides. By interplanting wild heliotrope (<u>Heliotropium europaeum</u>) within leguminous crops, weed populations have been reduced about 70 percent and the abundance of several insect pests reduced below an economic threshold as well (Putnam and Duke 1978). By introducing French and African marigolds in fields of certain crops, populations of nematodes were effectively controlled, and the germination of weeds such as morning glory, pigweed and Florida beggarweed was also partially inhibited (William 1981). Adaptations such as these provide a new context for agroecosystem management in which system stability depends on manipulating the ecological assemblage in fields to promote biotic interactions that benefit farmers (Altieri et al. 1983). In the following chapters, the most relevant ecological and agronomic features of five alternative production systems are described.

FARMING SYSTEMS RESEARCH

One of the key factors in modern agricultural development is the availability of an organized research, educational and extension infrastructure. Although in most countries there is such a network,

Table 4.2   Some agricultural technology approaches to reduce energy inputs into food production systems (expanded from Wittwer 1975)

Enhancement of photosynthetic efficiency

    Improvement of plant architecture for better light interception (i.e., leaves with vertical orientation)
    Genetic selection of varieties with greater efficiency (i.e., high leaf area index)
    Reduction or inhibition of photorespiration and/or night respiration
    Use of varieties of a more prolonged growth period
    Artificial enrichment with CO_2
    Hormonal stimulation of net photosynthesis
    Hormonal stimulation of crop senescence
    Genetic incorporation of C_4 or CAM mechanisms into C_3 crops
    Efficient planting patterns (orientation of rows N-S)
    Use of plastic mulches that reflect light back to underside of leaves

Environmental modification

    Wind modification with windbreaks and shelterbelts
    Frost control with windbreaks, heaters, fans and irrigation
    Control of soil temperatures through mulching or application of black charcoal and asphalt

Soil management

    Genetic selection of crops tolerant to nutritional deficiencies or toxicities
    Application of fertilizers at lower rates and increasing the efficiency of applied fertilizers
    Minimum or reduced tillage
    Use of manure, compost, cover crops and green manures
    Enhancement of biological N_2 fixation, and selection of bacteria able to fix N_2 in the rhizosphere of non-legume crops
    Use of Azolla in rice production
    Utilization of mycorrhizal associations
    Direct use of primary fertilizer sources (i.e., phosphoric rock)

Water management

    Drip irrigation
    Mulching, reduced tillage
    Control of stomata aperture with chemicals (i.e., PMA)
    Cover management for shade control
    Windbreaks
    Application of "required amounts" of water based on real soil water content

Insect pest management

    Preventative action: resistant varieties, manipulation of crop planting date, tillage and row spacing, crop rotation, improve field hygiene, use of attractants, pheromone traps, crop diversification, etc.
    Suppressive actions: Sterile male technique, sex-attractant pheromones, introduction, augmentation and conservation of parasites and predators, microbial and botanical insecticides, use of mechanical or fire removal, induction of behavioral changes, pesticidal controls when economic threshold is reached, etc.

Disease management

    Resistant varieties, crop rotations, use of sub-optimal fungicide doses, multilines or variety mixtures, biological control with antagonists, multiple cropping and reduced tillage

Weed management

    Design of competitive crop mixtures, rapid transplant of vigorous crop seedlings to weed-free bed, use of cover crops, narrow row spacings, crop rotation, keeping crop weed-free during critical competition period, mulching, cultivation regimes and allelopathy

Agronomic systems

    Multiple cropping systems: inter-cropping, strip-cropping, ratoon cropping, relay cropping, mixed cropping, etc.
    Use of cover crops in orchards and vineyards
    Sod strip intercropping and vegetable living-mulch system
    Agro-forestry systems
    Cropping systems analog to the natural secondary succession of the area

nowhere is it specifically directed to the needs and problems of alternative farmers. So far, most agricultural research has benefited those individuals with ready access to capital: large farmers and agribusinesses (Busch and Lacy 1983). In pest control, for example, an estimated 92 percent of the research effort is focused on the use of herbicides, and 55 percent and 89 percent, respectively, of the research is focused on applied pesticidal tactics in insect and pathogen control (Pimentel 1973). Also, virtually no researchers are examining alternative approaches to agricultural production, and therefore the wealth of information that can be extended to these farmers, even if a network existed, is very limited.

Clearly, the generation of technologies adapted to alternative farmers' needs must emerge from integrated studies of the natural and socioeconomic circumstances that influence their farming systems and dominate their responses to alternative technologies. Many circumstances may influence the type of cropping system or management practice a farmer chooses. Natural circumstances (climate, soil, pests, diseases) impose biological constraints on the crop system. On the other hand, many socioeconomic circumstances (transportation, capital, markets, labor, farm inputs, credit, technical assistance) affect the external environment that conditions farmers' decision making. By conducting multidisciplinary research in selected farmers' fields, and by analyzing the social, economic, technical and ecological constraints facing crop production in those farms, important feedback can be obtained about the farm conditions, management practices and farmers' needs. This information can then be incorporated into decisions on crop research conducive to the development of technologies adapted to alternative farmers' needs and resources.

Detailed descriptions of research methodologies attuned to actual conditions of traditional farmers of the developing world have recently become available (Harwood 1979, Hildebrand 1979, Byerlee et al. 1980, Zandstra et al. 1981, and Shaner et al. 1982). These methodologies emerged in response to the critics of internationally funded rural development, who charge that in the past programs lacked an understanding of the ecological and socioeconomic milieu in which they operated, excluded the small farmers as both collaborators and beneficiaries, and ineptly promoted inappropriate technology.

The new methodology, termed farming systems research, starts with an understanding of traditional agricultural systems. A multidisciplinary team gathers relevant information on the selected zone by analyzing background data from published or unpublished materials, conducting field surveys that include interviews with farmers and others knowledgeable about farm circumstances, and by field observations. From the survey, researchers can formulate hypotheses about why farmers use these particular practices (Table 4.3).

Field Procedures

Farming systems research follows a logical sequence of steps (Harwood, 1979):

1. Selection of the target area. Within regions, cropping systems and farmers' practices vary considerably. Sites with similar relative cropping patterns, agroclimatic characteristics and economic circumstances are selected by multidisciplinary teams, usually a group of economists, sociologists, agronomists and plant protectionists.

2. Description of the environment. Data are collected on climate, soil, topography, rainfall, drought, hydrology, temperature, day length, soil fertility, slope, insect pests, diseases and weeds.

3. Field surveys. The field survey includes a biophysical evaluation and a socioeconomic evaluation. The biophysical component entails (a) identifying land types at the site, (b) identifying existing crops, cropping patterns and systems (like rotations, polycultures), (c) describing cropping systems determinants, (d) describing farm types and the resource base at the site and (e) identifying farming system interactions, including those between crops and livestock, like complementarities such as crop residues

Table 4.3   A comparison between a typical top-down technology transfer approach (TOT) and a farming systems research approach (FSR) involving the farmers (after Chambers and Ghildyal 1985)

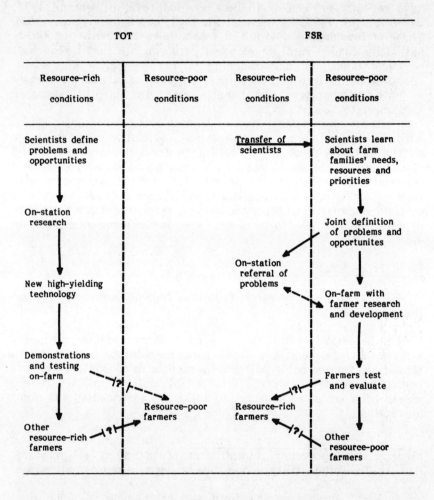

used for cattle feed and manure used as fertilizer. Researchers also describe the management practices by following the checklist described in Table 4.4.

The socioeconomic component of the survey analyzes the resources that go into cropping systems. Farm resources include (Shaner et al. 1982):

Table 4.4 A checklist of information on crop management practices to be recorded in each farm throughout the year (Byerlee et al. 1980)

Land Preparation
  Sequence of operations
  Timing of each operation in relation to rains
  Equipment used in each operation
  Variation in method with seasonal conditions

Planting
  Variety(ies) used
  Density and spacing
  Density and spacing of interplanted crops
  Time of planting in relation to rains, frosts, etc.
  Spread of planting dates
  Sequence of interplanting crops
  Method of planting (hills, broadcast, etc.)
  Method of covering seed
  Practices of replanting part or whole fields

Thinning
  Method and timing
  Target density
  Use of thinnings

Weeding
  Number of weedings
  Timing of each in relation to planting
  Equipment used in weeding
  Use of herbicides (type, rate, timing and method of application)
  Use of weeds

Fertilization
  Type of fertilizer(s) including organic
  Rate(s) of application
  Number and timing of applications
  Equipment used for application
  Method of application (e.g. broadcast, furrows, etc.)

Pest Control
  Method of control (type, rate, equipment)
  Timing of control

Irrigation
  Method of irrigation
  Frequency and timing of irrigation

Harvest
  Timing of harvest in relation to maturity
  Methods of harvesting
  Use of leaves and tops for animals
  Timing and method of picking leaves and tops
  Use of stalks

Post-Harvest
  Method of threshing/shelling
  Timing of threshing/shelling
  Method and quantity stored
  Disposal of produce (stored, sold, etc.)
  Use of crop in local foods

Seed Selection
  Time of selection
  Criteria for selection
  Special seed production or storage methods
  Seed treatment

LAND

Size of farm
Ownership
Permanency of use
Landlord/tenant relationships
Land quality (soil depth, texture and presence of toxic substances)
Terrain (slope, whether or not terraced)
Water availability (nearness of ponds or streams for livestock, irrigated or rainfed farming, dependability of supply)
Location: Access to markets and other services

LABOR

Members of the household or hired workers. Some relevant characteristics are:

>   Number, age and sex of family members and hired workers
>   Division of effort among the members and workers
>   Level of productivity and health
>   Division of time between on-farm and off-farm activities
>   Extent and nature of cooperative efforts
>   Other responsibilities that influence allocation of time and effort.

CAPITAL

Physical and financial assets that include:
Tools and equipment
Buildings and improvements to the land
Livestock and other assets capable of being sold to meet the farmers' needs
Cash from sale of crops, animal products, handicrafts, and from other sources

Access to credit

Cost and return analysis is used to measure the economic benefits of a new technology at the field level. However, a whole-farm analysis measures the economic benefits of both new and historical agricultural techniques in the context of all the economic activities, including other agricultural enterprises and household and off-farm operations. The major resources of land, labor and capital are all valued with due regard to their actual availability, and to the demand that exists for them at a particular time. Recommendations derived from cost/benefit data should be consistent with the farmers' desire to increase income and avoid risk, and with the scarcity of investment capital (Perrin et al. 1979).

TECHNOLOGY RESEARCH

Interpretation of the survey data allows researchers to plan experiments in farmers' fields. A group of farmers is selected to help design, test and evaluate experiments. A broad cross section of farmers must be included to avoid biasing the recommendations toward farmers of recognized ability. Experiments are designed to test particular component technologies (varietal selection, tillage and crop establishment methods, fertilization and pest management strategies) against farmers' current practices. A guiding principle is that the component technology should fit the resource limits of most farmers in the region, and should therefore be environmentally sound, socially acceptable and economically viable. However, the farmer should be able to decide freely which innovations are to be made on his or her land. A crop innovation may disturb the farmer's economic equilibrium and therefore requires a period of adjustment (Zandstra et al. 1981).

The productivity of the system should not be evaluated solely on the basis of crop yield per unit of land, but should incorporate the farmers' perspectives on productivity, emphasizing maximization of returns for the most limiting factor (such as labor, money, input). When the short-term economic performance of the cropping pattern is emphasized, the recommended technologies almost invariably further the use of chemical and mechanized technologies, typical of capital-intensive agriculture (Perelman, 1977).

Field trials usually include treatments that simulate and test the farmers' management level (which may not include any purchased material inputs), incorporate the technology thought to be optimal for the cropping pattern and evaluate a third level of inputs that are expected to produce still higher yields (Zandstra et al. 1981). Experiments must be repeated for two to five years to demonstrate the adaptation of new technologies to local conditions.

This methodology (Figure 4.1) could operate in the following way (Hart, 1978): Suppose maize and cassava are commonly intercropped in a humid tropical environment on farms of five hectares or less, with gross annual incomes between $500 and $1,000. A research team would analyze the agroecosystem in which the maize and cassava crop system is a subsystem, and the farm system in which the farmer's agroecosystem management plan operates. The team would then recommend modifications of the farmer's practices to improve annual income through increased crop yield. For example, the team might recommend a different maize variety, a change in the planting distance between the maize and cassava, more fertilizer or manure, or all three modifications. It is because of these successes that agricultural systems have great potential to serve as model systems in ecology. More active ecological research in agricultural ecosystems is clearly warranted.

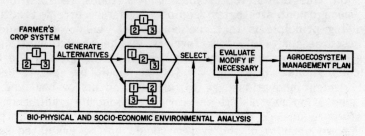

Figure 4.1 Methodological sequence in the modification of farmer's agroecosystem management plan (Hart 1978).

5
Designing Sustainable Agroecosystems

CHOOSING AN AGRICULTURAL SYSTEM

The first step in designing an agricultural system is to conceptualize it. Any concept of an agricultural system must include at least the following (Spedding 1975):

Purpose: Why the system is being established
Boundary: Where the system begins and ends
Context: The external environment in which the system operates
Components: The main constituents that form the system
Interactions: The relationships among the components
Inputs: Items used by the system that come from outside it
Resources: Components already within the system that are used in its functioning
Products or performance: The primary desired outputs
By-products: Useful but incidental outputs.

The second step is to match the needs of the conceptualized system as closely as possible with the local constraints, conditions and available resources (Spedding 1975). The considerations that determine the feasibility, profitability, practicality and preferences are summarized in Table 5.1.

Clearly, environments will differ both in their resources and constraints, and in the extent to which these can be modified. Resource needs can also be modified somewhat but all such modifications involve some cost. In general, systems based on annual staple crops require less investment and environmental modification than specialty vegetable or fruit crop systems (Table 5.2).

Table 5.1 Factors affecting the choice of farming systems (after Spedding 1975)

Ecological factors	Infra-structural features	External economic constraints	Internal operational factors	Personal acceptance
Climatic	Land tenure	Markets	Farm size	Personal preferences
Soil	Water supply	Communications	Labor availability	
Biological	Power supply	Credit availability		

Elements of Sustainability

The basic tenets of a sustainable agroecosystem are conservation of renewable resources, adaptation of the crop to the environment and maintenance of a high but sustainable level of productivity. To emphasize long-term ecological sustainability rather than short-term productivity, the system must:

o   Reduce energy and resource use
o   Employ production methods that restore homeostatic mechanisms conducive to community stability, optimize the rate of turnover and recycling of matter and nutrients, maximize the multiple-use capacity of the landscape and ensure an efficient energy flow
o   Encourage local production of food items adapted to the natural and socioeconomic setting
o   Reduce costs and increase the efficiency and economic viability of small and medium-sized farms, thereby promoting a diverse, potentially resilient agricultural system.

Sustainability can best be achieved through an understanding of the four subsystems of agriculture (Raeburn 1984):

1. Biological: Plants and animals and the biological effects of physical and chemical factors (climate, soil) and of management activities (irrigation, fertilization, tillage) on plant and animal performance.

Table 5.2 Some factors favoring success in modern agroecosystems (after Thorne and Thorne 1979)

	Requirements	
Factors	Field Crops	Vegetable, fruits or special crops
Size of farm	Variable, small to large if mechanized for harvesting	Variable, small to medium
Climate	Limits, kinds and varieties of crops	A restriction for many specific crops, especially frosts
Soil	Classes I to III depending on soil conservation practices	Class I, but many crops have special requirements (flat soils, high fertility, etc.)
Water	Good water supply, can adapt to some arid conditions	Needs good water supply
Labor requirements	Varied	Generally high
Specialized labor	Medium	High for some crops
Capital investment machinery, buildings	Varied	Generally high
Fertilizer requirement	High, especially nitrogen	High and varied. Many require micronutrients
Pest control	Varied depending on plant diversity	High for some crops demanding high cosmetic quality
Use of crop rotations	Varied	Varied, lacking in fruit crops

2. Work: The physical tasks of agriculture and how they can be achieved by combining labor, skills, machinery and energy.

3. Farm economics: The cost of production and the prices of crops being raised, quantities produced and used, risks and all other determinants of farm income.

4. Socioeconomic: Markets for farm products, land use rights, labor, machinery, fuel, inputs, credit, taxation, research, technical assistance, etc.

Models for Agroecosystem Design

The physiological limits of crops—the carrying capacity of the habitat and the external costs of enhancing production—put a ceiling on potential productivity. This point is the "management equilibrium" (Lewis 1959) at which the ecosystem, in dynamic equilibrium with environmental and management factors, produces a sustained yield. The characteristics of this equilibrium will vary with different crops, geographical areas and management objectives, so they will be highly site-specific. However, general guidelines for designing balanced and well-adapted cropping systems may be gleaned from the study of structural and functional features of the natural or seminatural ecosystem remaining in the area where agriculture is being practiced. Four major sources of "natural" information can be explored:

Primary production. Depending on climatic and edaphic factors, each area is characterized by a type of vegetation with a specific production capacity. A natural grassland area (with a standing crop value of 6600 g/m^2 is not able to support a forest 26,000 grams per square meter) unless subsidies are added to the system. It follows, then, that if a natural grassland needs to be transformed into an agricultural system, it should be replaced by cereals rather than orchards.

Land use capability. Soils have been classified into eight land use capability groups, each determined by physiochemical factors, such as slope or water availability (Vink 1975). According to this classification, soils of class I and II are highly fertile, have good texture and permeability and are deep and erosion-resistant; in short, they are suitable for many types of crops. However, when trees and shrubs are replaced by wheat on hillsides (i.e., class VI soil), yields decline progressively and the soil becomes badly eroded (Gasto and Gasto 1970). In determining the suitability of a tract of land for a certain agricultural use, it is important to consider qualities such as availability of water, nutrients and oxygen; soil texture and depth; salinization and/or alkalinization; possibilities for mechanization; and resistance to erosion (Vink 1975). Figure 5.1 shows the relationship between USDA land capability classes and the intensity with which each class can be used safely.

Vegetational patterns. The vegetation of a natural ecosystem can be used as an architectural and botanical model for designing and structuring an agroecosystem to replace it. The study of productivity, species composition, efficiency of resource use, resistance to pests and leaf area distribution in natural plant communities is important for building agroecosystems that mimic the structure and function of natural successional ecosystems (Ewell 1986). In the humid tropical lowlands, Ewel argues that constructing forest-like agroecosystems that imitate successional vegetation is the only means of constructing a sustainable agriculture. Such agroecosystems would exhibit low requirements for fertilizer, high use of available nutrients and high protection from pests.

This succession analog method requires a detailed description

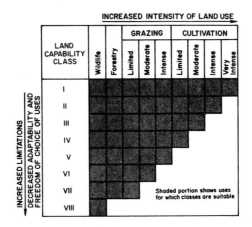

Figure 5.1 Relationship between land capability classification classes and the intensity with which each class can be used (Vink 1975).

of a natural ecosystem in a specific environment and the botanical characterization of all potential crop components. When this information is available, the first step is to find crop plants that are structurally and functionally similar to the plants of the natural ecosystem. The spatial and chronological arrangement of the plants in the natural ecosystem are then used to design an analogous crop system (Hart 1978). In Costa Rica, Ewel et al. (1984) conducted spatial and temporal replacements of wild species by botanically/structurally/ecologically similar cultivars. Thus, successional members of the natural system such as Heliconia spp., cucurbitaceous vines, Ipomoea spp., legume vines, shrubs, grasses and small trees were replaced by plantain, squash varieties, yams, By years two and three, fast-growing tree crops (Brazil nuts, peach, palm, rosewood) may form an additional stratum, thus maintaining continuous crop cover, avoiding site degradation and nutrient leaching and providing crop yields throughout the year (Uhl and Murphy 1981).

Gasto (1980) designed a similar conversion system in the Mediterranean matorral of central Chile. Matorral vegetation consists of shrubs (dominated by Acacia caven) and an understory of mixed grasses. Successful sheep pastures were developed by replacing the natural shrub layer with Atriplex spp. shrubs, a food source for the animals. Thus, species composition was altered, but the structural profile was left intact.

Knowledge of local farming practices. In most rural areas farmers have been cultivating for decades. Some have failed and others have succeeded in developing cropping systems adapted to local conditions. Despite the onrush of modernization and economic change, a few traditional agricultural management systems survive. These systems exhibit important elements of sustainability; namely, they are well adapted to their environment, they rely on local resources, they are small-scale and decentralized and they conserve natural resources. At the field level, the traditional polycultures often parallel natural plant communities by containing:

o Genetic diversity in plant species
o Complex trophic relationships among crops, weeds, insects and pathogens
o Relatively closed nutrient cycles, with many crop nutrient requirements supplied by rotations, fallow or manure
o Year-round vegetative cover of soil
o Efficient use of water, sunlight and soil
o Low risk of complete loss of crops, due to diversity
o High level of production stability due to compensation by the various components when one component fails.

Thus, although tropical small farmers with little capital or institutional support have been confined to farming low quality, marginal soils, their systems provide valuable information for the development of yield-sustaining systems.

CHOOSING A CROPPING SYSTEM

Crop production systems include both cropping systems and the associated crop production practices and technologies used to raise crops. Cropping systems may consist of a monoculture of one continuous crop or formal sequences of crops repeated in an orderly pattern to constitute a rotation. They may also include flexible arrangements of one or more crops in time and space (intercropping, relay cropping), and intensive successions of crops within single years or even within seasons. Cropping systems vary greatly with differences in soil and climate and with local economic and social systems.

Crop growth and performance are subject to environmental conditions (topography, rainfall, soil texture and fertility) and management conditions (planting time, weeding). Before designing new cropping systems in an area already being farmed, the existing systems must be described in terms of rainfall and temperature (Beets 1982). A useful start is a simple climatic diagram with the months on the X axis and the average temperature (degrees

Celsius) on the left side of the Y axis and the average precipitation (mm) on the right side of the Y axis, maintaining the relationship of one degree Celsius to 2 mm of precipitation. This relationship roughly approximates evaporation; when the precipitation curve is below that of the temperature curve it denotes a period of drought. When precipitation is above temperature there is enough moisture for crop growth. Thus, in the Central Plateau of Mexico, an analysis of this diagram (Figure 5.2) depicts four periods of great agronomic importance:

1. Low risk of frost at the end of spring
2. Beginning of rains
3. Average growing period
4. First autumn frosts

Several agronomic considerations are involved in developing a cropping system (Thorne and Thorne 1979). Cropping systems should be devised to provide high photosynthetic capabilities for as much of the year as practical. In intercropped or mixed crops, plant height, the shape and angle of leaves and the rate of growth and time period required to reach maturity are important characteristics that determine photosynthetic efficiency. There are several ways to combine crop plants to maximize solar radiation, such as by combining species of different phenologies, that reach maximum photosynthesis at different radiation levels, or that have roots that exploit different parts of the soil.

A major goal should be to maximize annual crop production or net economic gains per unit area of land. Thus, two short-season crops may provide greater total yields than one long-season crop. Decisions as to crop intensities must be based on the best available evidence for each combination of conditions. To promote sustained high yields and profits, cropping systems should be designed to maintain soil organic matter and tilth; to reduce the incidence of weeds, insects and diseases; to help keep plant nutrients in balance; to conserve water; and to minimize soil erosion.

To use water and nutrients effectively, roots should form an active and extensive network throughout the soil. Good crop combinations have compatible root systems that permeate the soil to a depth of 25 to 30 cm with some roots extending much deeper. One crop may root deeper than another to good advantage.

CROP CHARACTERISTICS AND CROPPING PATTERNS

Biological and agronomic characteristics of crop plants are important in selecting crops for any given situation and in determining appropriate farming practices. These characteristics

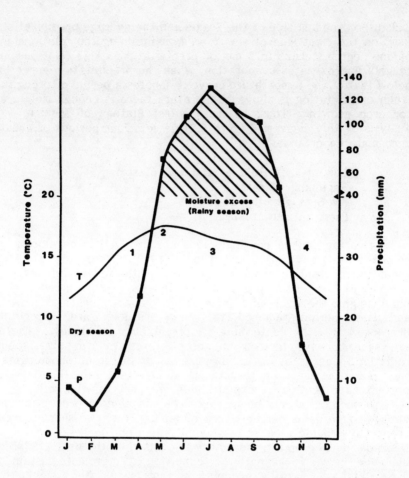

Figure 5.2 Climatic diagram of the Central Plateau of Mexico (Williams 1985)

can be summarized as follows (Thorne and Thorne 1979):

  Growing period. The number of days required between date of plant emergence and maturity is important both in determining the correct climatic zone for a specific crop and in fitting a particular cultivar into a multiple cropping system.

  Photoperiodism. For many plants the length of night (darkness) rather than length of daylight is critical to initiate flowering, tillering or dormancy. Short-day plants require prolonged daily darkness to induce flowering, and long-day plants initiate flowering when nights are relatively short. Some plants are day-neutral, and develop without regard to daylength. In some plants the change in daylength may be important to induce changes in development. Increasing daylength may help initiate flowering, while in the fall the advent of shorter days may promote fruiting, maturity or dormancy.

Growth habits. The growth habits of crop plants are important in determining production and management practices. Dwarf varieties are generally preferred over their tall counterparts because of their upright growth habit, greater ease of harvesting by machine, reduced likelihood of lodging, earlier fruiting and frequently higher harvest index. Bush varieties are preferred over vines because they have many relatively dwarfed branches that bear fruit uniformly.

Root systems. Two types of root systems are common to crop plants: fibrous roots and tap roots. Fibrous roots permeate the soil and hold soil particles together. Grasses, for example, promote good soil structure and help protect soils against erosion. Tap-rooted crops are those with roots commonly harvested for food or feed, such as sugar beets, mangels, carrots and turnips. Tap-rooted plants tend to be deep-rooted, such as alfalfa and trees. Deep-rooted plants maximize the upward flow of both soluble and less soluble nutrients.

In most crop plants the major volume of roots is in the upper 30 cm of soil. However, the depth of intensive rooting is affected by soil moisture, texture, compaction and aeration, and the supply of available plant nutrients.

6

Traditional Agriculture

About 60 percent of the world's cultivated land is still farmed by traditional and subsistence methods (Ruthenberg 1971). This type of agriculture has benefited from centuries of cultural and biological evolution that has adapted it to local conditions (Egger 1981). Thus, small farmers have developed and/or inherited complex farming systems that have helped them meet their subsistence needs for centuries, even under adverse environmental conditions (on marginal soils, in drought or flood-prone areas, with scarce resources) without depending on mechanization or chemical fertilizers and pesticides. Generally these farming systems consist of a combination of production and consumption activities (Figure 6.1).

Most small farmers have employed practices designed to optimize productivity in the long term rather than maximize it in the short term (Gliessman et al. 1981). Inputs characteristically originate in the immediate region and farm work is performed by humans or animals that are fueled from local sources (Figure 6.2). Working within these energy and spatial constraints, small farmers have learned to recognize and use locally available resources (Wilken 1977). Traditional farmers are much more innovative than many agriculturalists believe. Many scientists in developed countries are beginning to show interest in traditional agriculture, especially in small-scale mixed crop systems, as they search for ways to remedy deficiencies in modern agriculture. This transfer of learning must occur rapidly, however, or this wealth of practical knowledge will be lost forever.

ECOLOGICAL FEATURES OF TRADITIONAL AGRICULTURE

As more research is conducted, many farming practices once regarded as primitive or misguided are being recognized as

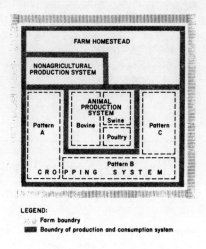

Figure 6.1 Scheme of a small farming system with four production consumption systems (Zandstra et al. 1981).

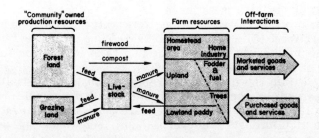

Figure 6.2 Conceptual model of the production system of a Nepalese hill farm (Harwood 1979).

sophisticated and appropriate. Confronted with specific problems of slope, flooding, droughts, pests, diseases and low soil fertility, small farmers throughout the world have developed unique management systems to overcome these constraints (Table 6.1). Traditional agriculturalists generally have met the environmental requirements of their food-producing systems by concentrating on a few principles and processes (Knight 1980):

Spatial and temporal diversity and continuity: Multiple cropping designs are adopted to ensure constant food production and vegetation cover for soil protection. By ensuring a regular and varied food supply, a diverse and nutritionally adequate diet is assured. Extended crop harvest reduces the necessity for storage, often hazardous in rainy climates. A continuous sequence of crops also maintains biotic relationships (predator/prey complexes, nitrogen fixing) that may benefit the farmer.

Optimal use of space and resources: Assemblages of plants with different growth habits, canopies and root structures allows for better use of environmental inputs such as nutrients, water and solar radiation. Crop mixtures make fullest use of a particular environment. In complex agroforestry systems, crops can be grown underneath tree canopies if enough light gets through.

Recycling of nutrients: Small farmers sustain soil fertility by maintaining closed cycles of nutrients, energy, water and wastes. Thus, many farmers enrich their soils by collecting nutrient materials (such as manure and forest litter) from outside their fields, adopting fallow or rotational systems or including legumes in their intercropping patterns.

Water conservation: In rainfed areas, the rainfall pattern is the main cropping system determinant, and farmers use cropping patterns adapted to the amount and distribution of rainfall. Thus, in areas with little moisture, farmers prefer drought-tolerant crops (like Cajanus, sweet potato, cassava, millet and sorghum), and management techniques emphasize soil cover (such as mulching) to avoid evaporation and runoff. Where precipitation is more than 1,500 mm/year, most cropping systems are based on rice. Under constant flooding conditions, instead of investing in costly drainage systems, farmers develop integrated agriculture-/aquaculture systems, such as the chinampas of Central Mexico.

Control of succession and protection of crops: Farmers have developed a number of strategies to cope with competition from undesirable organisms. Crop species and variety mixtures provide insurance against catastrophic attacks from insect pests or disease.

Table 6.1 Some examples of soil, space, water and vegetation management systems used by traditional agriculturalists throughout the world (after Klee 1980)

Environmental constraint	Objective	Recommended practice
Limited space	Maximize use of environmental resources and land	Intercropping, agroforestry, multi-story cropping, home gardens, altitudinal crop zonation, farm fragmentation, rotations
Steep slopes	Control erosion, conserve water	Terracing, contour farming, living and dead barriers, mulching, leveling, continuous crop and/or fallow cover, stone walls
Marginal soil fertility	Sustain soil fertility and recycle organic matter	Natural or improved fallow, crop rotations and intercropping with legumes, litter gathering, composing, manuring, green manuring, grazing animals in fallow fields, night soil and household refuse, mounding with hoe, ant hills as source of fertilizer, use of alluvial deposits, use of aquatic weeds and muck, alley cropping with legumes, plowed leaves, branches and other debris, burning vegetation, etc.
Flooding or excess water	Integrate agriculture with water supply	Raised field agriculture (chinampas, tablones), ditched fields, diking, etc.
Excess water	Channel/direct available water	Control floodwater with canals and checkdams. Sunken fields dug down to groundwater level. Splash irrigation. Canal irrigation fed from ponded groundwater, wells, lakes, reservoirs
Unreliable rainfall	Best use of available moisture	Use of drought-tolerant crop species and varieties, mulching, weather indicators, mixed cropping using end of rainy season, crops with short growing periods

Temperature or radiation extremes	Ameliorate microclimate	Shade reduction or enhancement; plant spacings; thinning; shade-tolerant crops; increased plant densities; mulching; wind management with hedges, fences, tree rows; weeding; shallow plowing; minimum tillage; intercropping; agroforestry; alley-cropping, etc.
Pest incidence (invertebrates, vertebrates)	Protect crops, minimize pest populations	Overplanting, allowing some pest damage, crop watching, hedging or fencing, use of resistant varieties, mixed cropping, enhancement of natural enemies, hunting, picking, use of poisons, repellants, planting in times of low pest potential

Crop canopies can effectively suppress weed growth and minimize the need for weed control. In addition, cultural practices such as mulching, changes in planting times and durability, use of resistant varieties and use of botanical insecticides and/or repellants can minimize pest interference.

ADVANTAGES OF CROP DIVERSITY

Perhaps one of the most striking features of traditional farming systems in most developing countries is the degree of crop diversity both in time and space. This diversity is achieved through multiple cropping systems, or polycultures. For example, in the Latin American tropics, 60 percent of the corn is grown intercropped (see Chapter 9). Similarly, in Nigeria 98 percent of the cowpea, the country's most important legume, is grown in association with other crops.

Polyculture is a traditional strategy to promote diet diversity, income generation, production stability, minimization of risk, reduced insect and disease incidence, efficient use of labor, intensification of production with limited resources and maximization of returns under low levels of technology (Francis et al. 1976, Harwood 1979). Polyculture sytems offer many advantages over the monoculture agriculture practiced in modern countries, as follows (Ruthenberg 1971, Altieri 1983, Francis 1986):

Yields. Total yields per hectare are often higher than sole-crop yields, even when yields of individual components are reduced. This yield advantage is usually expressed as the land equivalent ratio (LER), which expresses the monoculture land area required to produce the same amount as one hectare of polyculture, using the same plant population. If the LER is greater than one, the polyculture overyields. Most corn/bean dicultures and corn/bean/squash tricultures studied are examples of overyielding polycultures.

Efficient use of resources. Mixtures result in more efficient use of light, water and nutrients by plants of different height, canopy structure and nutrient requirements. There is some indication that long-duration intercrop combinations have an advantage when nutrients are limited. Thus, in polycultures combining perennial and annual crops, the minerals lost by annuals are rapidly taken up by perennials. On the other hand, the nutrient-robbing propensity of some crops is counteracted by the enriching addition of organic matter to the soil by other crops (like legumes) in the mixture.

Nitrogen availability. In cereal/legume mixtures, fixed nitrogen from the legume is available to the cereal, thereby improving the nutritional quality of the mixture. Corn and beans complement each other in essential amino acids.

Reduction of diseases and pests. Diseases and pests may not spread as rapidly in mixtures because of differential susceptibility to the pests and pathogens and because of enhanced abundance and efficiency of natural enemies. In Southeast Asia, for example, maize grown in rows two to three meters apart, intercropped with soybeans, groundnuts, upland rice or mungbean, suffers relatively little from downy mildew, normally a major maize disease. Similarly, in Costa Rica, cowpea mosaic and chlorotic viruses occurred at lower levels in cowpea intercropped with cassava than in cowpea monocultures (Altieri and Liebman 1986). Diversified crop systems can increase opportunities for natural enemies, and consequently improve biological pest control. Two-thirds of the studies dealing with the effects of crop diversity on insect pests showed that pestiferous insects decreased in the diversified system when compared with the corresponding monoculture. In many cases this was due to the abundance and efficiency of natural enemies. Cabbage aphids, flea beetles, fall armyworm, imported cabbage butterfly, leaf beetles, diamondback moth and corn earworm are all insect pests that can be regulated with specific crop mixes (Altieri and Letourneau 1982).

Weed suppression. The shading provided by complex crop canopies helps to suppress weeds, thereby reducing the need and cost of weed control. In the Philippines, shade-sensitive weeds such as nutsedge and Imperata cylindrica may be eliminated entirely by a combination like maize/mungbean, which intercepts 90 percent of the light after 50 days of growth.

Insurance against crop failure. Polycultures provide insurance against crop failure, especially in areas subject to frosts, floods or droughts. Thus, when one of the crops in a combination is damaged early in the growing season, the other crops may compensate for the loss. For example, in the highlands of Tlaxcala, Mexico, farmers intercrop corn with fava beans, because fava beans survive frosts, whereas corn does not.

Other advantages. Polycultures provide effective soil cover and reduce the loss of soil moisture. They enhance opportunities for marketing, ensuring a steady supply of a range of products without much investment in storage, thus increasing the marketing success. Mixtures spread labor costs more evenly throughout the cropping season, and usually give higher gross returns per unit of labor employed, especially during periods of labor scarcity. Polycultures also can improve the local diet; 500 grams of maize and 100 grams of black beans per day provide about 2118 calories and 68 grams of protein daily.

EXAMPLES OF TRADITIONAL FARMING SYSTEMS

Paddy Rice Culture in Southeast Asia

Beneath the simple structure of the rice paddy monoculture (sawah) lies a complex system of built-in natural controls and genetic crop diversity (King 1927). Although these systems are more prevalent in Southeast Asia, upland rice farmers in the Latin American tropics also grow a number of photoperiod-sensitive rice varieties adapted to differing environmental conditions. These farmers regularly exchange seed with their neighbors because they observe that any one variety begins to suffer from pest problems if grown continuously on the same land for several years. The temporal, spatial and genetic diversity resulting from farm-to-farm variations in cropping systems confers at least partial resistance to pest attack. Depending on the degree of diversity, food web interactions among the insect pests of rice and their numerous natural enemies in paddy fields can become very complex, often resulting in low but stable insect populations (Matteson et al. 1984).

The rice ecosystem, where it has existed over a long period, also includes diverse animal species. Some farmers allow flocks of domestic ducks to forage for insects and weeds in the paddies. Many farmers allow aquatic weeds, which they harvest for food (Datta and Banerjee 1978). Frequently one finds paddies where farmers have introduced a few pairs of prolific fish (such as common carp, Sarotherdon mossambicus, swamp fish species). When the water is drained off to harvest the rice, the fish move to troughs or tanks dug in the corners of fields and are then harvested.

The techniques used for rice/fish culture differ considerably from country to country and from region to region. In general, exploitation of rice field fisheries may be classified as captural or cultural (Pullin and Shehadeh 1980). In the captural system wild fish populate and reproduce in the flooded rice fields and are harvested at the end of the rice-growing season. Captural systems occupy a far greater area than cultural systems and are important in all the rice-growing areas of Southeast Asia. In the cultural system the rice field is stocked with fish. This system may be further differentiated into a concurrent culture, in which fish are reared concurrently with the rice crop, and a rotation culture, in which fish and rice are grown alternately. Fish can also be cultured as an intermediate crop between two rice crops (Ardiwinata 1957).

Traditional paddy rice growers usually produce only one rice crop each year during the wet season, even when irrigation water is readily available. This practice is partly an attempt to avoid damage by rice stem borers. For the remainder of the year the land may lie fallow and be grazed by domestic animals. This annual fallow, along with the dung dropped by the grazing animals and the weeds and stubble plowed into the soil, will usually sustain acceptable rice yields (Webster and Wilson 1980).

Alternatively, farmers may follow rice with other annual crops in the same year where adequate rainfall or irrigation water is avilable. Planting alternate rows of cereals and legumes is common, as farmers believe it uses the soil resources more efficiently. Well-rotted composts and manures are applied to the land to provide nutrients for the growing crops. Sowing cowpeas or mung beans into standing rice stubble reduces damage by bean flies, thrips and leafhoppers, by interfering with their ability to find their host (Matteson et al. 1984).

The micro-environment of the sawah also helps the wet-rice cultivator to produce constant crop yields from the same field year after year. First, the water-covered sawah is protected from high temperatures and the direct impact of rain and high winds, thus reducing soil erosion. Second, the high water table reduces the vertical movement of water, thus limiting nutrient leaching. Third, both floods and irrigation water bring silt in suspension and other plant nutrients in solution, renewing soil fertility each year. Fourth, the water in the sawahs contains Azolla spp. (a symbiotic association of blue-green algae and fern), which promotes the fixation of nitrogen, adding up to 50 kg per hectare of nitrogen.

Javanese Traditional Agriculture

In Java, Indonesia, many traditional agricultural systems combine crops and/or animals with tree crops or forest plants.

Some of these are agroforestry systems and can be grouped into two major types (Marten 1986):

Talun-kebun. This is an indigenous Sundanese agricultural system that appears to have derived from shifting cultivation. It usually consists of three stages—kebun, kebun campuran and talun—each of which serves a different function. In the kebun, the first stage, a mixture of annual crops is usually planted. This stage is economically valuable since most of the crops are sold for cash. After two years, tree seedlings have begun to grow in the field and there is less space for annual crops. At this point the kebun gradually evolves into a kebun-campuran, where annuals are mixed with half-grown perennials. The economic value of this stage is not as high, but it has high biophysical value, as it promotes soil and water conservation. After harvesting the annuals, the field is usually abandoned for two to three years to become dominated by perennials. This third stage is known as talun, and has both economic and biophysical values.

After the forest is cleared, the land can be planted to huma (dryland rice) or sawah (wet rice paddy), depending on whether irrigation water is available. Alternatively, the land can be turned directly into kebun by planting a mixture of annual crops. In some areas kebun is developed after harvesting the huma by following the dryland rice with annual field crops. If the kebun is planted with tree crops or bamboo, it becomes kebun campuran (mixed garden), which after several years will be dominated by perennials and become talun (perennial crop garden). It is not uncommon to find talun-kebun composed of up to 112 species of plants. Of these plants about 42 percent provide for building materials and fuelwood, 18 percent are fruit trees, 14 percent are vegetables and the remainder constitute ornamentals, medicinal plants, spices and cash crops.

Pekarangan (home garden). The pekarangan is an integrated system of people, plants and animals with definite boundaries and a mixture of annual crops, perennial crops and animals surrounding a house. A talun-kebun is converted into a pekarangan when a house is built upon it. Instead of clearing the trees to cultivate field crops as in talun-kebun, the home garden trees are kept as a permanant source of shade for the house and the area around it, and field crops in the home garden are planted beneath the trees.

A typical home garden has a vertical structure from year to year, though there may be some seasonal variation. The number of species and individuals is highest in the lowest story and decreases with height. The lowest story (less than one meter in height) is dominated by food plants like spices, vegetables, sweet potato, taro, Xanthosoma, chili pepper, eggplant and languas. The next layer (one to two meters in height) is also dominated by food plants, such as ganyong (Canna edulis), Xanthosoma, cassava, and gembili (Dioscorea esculenta). The next story (two to five meters) is dominated by bananas, papayas, and other fruit trees. The five- to ten-meter layer is also dominated by fruit trees, for example soursop, jack fruit, pisitan (Lansium domesticum), guava, mountain apple or other cash crops, such as cloves. The top layer (10 meters) is dominated by coconut trees and trees for wood production, like Albizzia and Parkia. The overall effect is a vertical structure similar to a natural forest, a structure that optimizes the use of space and sunlight. The most common plants in the pekarangan are cassava (Manihot esculenta) and ganyong (Canna edulis). Both have a high calorie content and are important as rice substitutes.

There are definite groupings of plants in the home garden. For example, wherever gadung is found, petai (Parkia speciosa), kemlakian and rambutan, possibly guava (Psidium guajava) and suweg (Amorphophalus campanulatus) will probably also be present.

An important plant association consists of rambutan (Nephelium lappaceum), kelor, (Moringa pterygosperma), rose (Rosa hybrida), mangkokan (Polyscias scutelaria), gadung (Dioscorea hispida) and grapefruit (Citrus grandis). Each of the plants in this association provides the farmer with something useful. Rambutan fruit is sold and eaten; kelor is used as a vegetable and is also believed to be a magical plant; rose is grown for pleasure; mangkokan is grown as an aesthetic plant and is used occasionally for hedges and hair tonic; gadung is a food plant that can also be used as a weather indicator because the rainy season usually begins a short time after its leaves start to grow; grapefruit has a similar function, and when its fruits start to grow the season of annual plant cultivation begins. These weather and planting-time indicators are important; many farmers believe agricultural failures are due mainly to improper planting times.

Shifting Cultivation

Shifting cultivation is also called slash-and-burn or swidden agriculture. These systems involve a few years of cultivation alternating with several years of fallow to regenerate soil fertility. Typically there are three types of fallow: forest fallow (20 to 25 years), bush fallow (six to 10 years) and grass fallow (less than five years).

Within the tropics, shifting cultivation is most important in Africa. In Asia and tropical America it is practiced by disadvantaged people in remote rural areas where the lack of roads precludes the development of markets for cash crops. In South and Southeast Asia, about 50 million people are shifting cultivators, cropping 10 to 18 million hectares each year. With the gradual development of rice cultivation in lowland areas, shifting cultivation has retreated to hilly areas unsuitable for paddy. In tropical America, shifting cultivation was practiced before 1,000 B.C. It is based on corn, beans and squash in the drier tropical areas of Mexico, and on tubers, cassava and sweet potatoes in the wetter lowlands (Norman, M. 1979).

The features of shifting cultivation include (Grigg 1974):

o   The size and number of plots managed by each family varies with the soil fertility, population density, length of the fallow and degree of commercialization.
o   It may or may not require a shift of domicile.
o   Land tenure is usually communal, and most farmers have cooperative arrangements to work the land, particularly to clear the vegetation.
o   Methods of cultivation are based on human and animal power, characterized by hand tools.
o   Farm livestock play a minor role.
o   Little cultivation and management are done once the crops are sown.
o   Generally soil fertility is maintained with some animal manure, but mostly with the nutrients provided by the ash and decomposing vegetation.

It is common in shifting cultivation to cut a parcel of forest and burn the area to release nutrients and eliminate weeds. A mixture of short-term crops, sometimes followed by perennials, is grown until the soil loses its fertility and competition from successional plant species is severe. Then the farmer prepares a new field and the old one returns to long-term fallow. During the fallow period, large quantities of nutrients are stored in the plant biomass. These nutrients are released when the fallow vegetation is burned to clear the land for the next cropping cycle (Ruthenberg 1971). Where land is abundant and resources scarce, it is generally agreed this is an efficient and stable system that has sustained farm families for many generations. Due to recent population pressure and factors like weed growth and declining soil fertility, the fallow cycle has been reduced from a more favorable 20 to 30 years to a period as short as five years, leading in many cases to soil losses and nutrient depletion.

Although there is generally a random generation of species during the fallow periods, in certain parts of the humid tropics farmers intentionally retain certain species such as Acioa baterii, Anthonata macrophylla and Alchornia sp. The small trees are only trimmed and the big branches are left for staking crops. The cut tops are spread on the soils and burned. Thus, the bush fallow functions doubly to provide staking materials and recycle nutrients (Nye and Greenland 1961).

It has been speculated that bush fallows are potentially valuable in controlling insects. The great diversity of crops grown simultaneously in shifting cultivation helps prevent pest buildup on the comparatively isolated plants of each species. Increased parasitoid and predator populations, decreased colonization and reproduction of pests, chemical repellency or masking, feeding inhibition from non-host plants, prevention of pest movement or stimulation of pest emigration and optimal synchrony between pests and their natural enemies are presumably important temporal and spatial factors in regulating pests in polycultures. Shade from forest fragments still standing in new fields, coupled with a partial canopy of fruit, nut, firewood, medicinal and/or lumber tree species, reduces shade-intolerant weed populations and provides alternative hosts for beneficial (or sometimes detrimental) insects. Clearing comparatively small plots in a matrix of secondary forest vegetation permits easy migration of natural control agents from the surrounding jungle (Matteson et al. 1984).

The nkomanjila system of the Nyhia shifting cultivators. This is a typical shifting cultivation system and involves a cycle of cutting woodland, burning, cropping and fallow (King, 1978). The Nyiha cultivators prefer fully regrown or virgin woodland composed of specific trees, such as Brachystegia spp. and Acacia macrothyrsa.

In burning the nkomanjila, cut wood is stacked around tree trunks and burned just before the rainy season. If much unburned material remains, it is gathered and reburned. After one month, during which other fields are prepared, the crops are planted. Before planting, ash from the burned trees is spread with hoes evenly through the field, and weeds are hoed into the soil. Seeds are broadcast and lightly hoed.

The nkomanjila must be weeded, usually once but sometimes twice. Women do the weeding and harvesting, while men do the cutting, burning and some of the initial hoeing. After harvesting, the crop is dried in the sun and then stored.

Nkomanjila crops include finger millet, perennial sorghum, pulses (including pigeon peas, lima beans and cowpeas) and cucurbits (including pumpkins and gourds) in intercropping patterns.

The standard crop sequence in nkomanjila is the finger millet/sorghum complex the first year, followed by sorghum ratoons or suckers the second year. A portion of the first-year field may be planted to an early maturing variety of finger millet. The second year of sorghum ratoons (lisala) is virtually untended except for harvesting. Traditionally, the cropping sequence ended here and the field was abandoned to fallow. Today, however, so much acreage is needed for food production that these long crop sequences are no longer possible. The basic two-year nkomanjila sequence may be repeated. Alternatively, if the finger millet yield the first year is good, the basic pattern may be reinitiated in the second year. Once the nkomanjila is abandoned (which may be due to weed growth or lower soil fertility), the land is allowed to rest about five to seven years.

Given current population densities in Africa, the nkomanjila system is no longer viable since farmers can no longer afford long fallow periods. As a result of frequent cultivation and burning, a cultivation system involving a grass-dominated fallow has replaced the woodland-dominated fallow.

The Nkule system. The nkule system is the grassland alternative to nkomanjila (King 1978). Techniques used in nkule cultivation can be applied both to upland grass communities, resulting in fields known as nkule, and at higher elevations, where the fields are called ihombe. Indicators for the nkule method include tall grasses of the Hyparrhenia genus and Trachypogon spicatus.

The distinctive feature of the nkule system is that turf and soil are mounded over grass, which is then burned under the mound. Maize and cucurbits are planted under the mound. In December the mounds around these crops are hoed down. Ash and burned soil are then spread and finger millet is sowed. The field requires two weedings, one in the course of hoeing down the mounds and preparing the seedbed, and a second during the growing season. The finger millet crop is harvested and stored as under the nkomanjila system. In ihombe fields, mounds are made and burned, but finger millet is the only crop planted after the mounds have been spread.

Burning both vegetative matter and surface soil is important in the nkule system. Usually, fallow grassland is plowed during the dry season to break the sod, which is then hoed into mounds. Cow dung is put on the windward side of the mound and set afire. The soil and turf of the mound is slowly hoed over the burning dung until all vegetative matter has been burned.

The important difference between nkule and ihombe fields is that in the nkule crops are virtually always growing, while the ihombe is used for just one year and fallowed at least three years. In upland nkule fields, the crop sequences are as varied as after nkomanjila. An upland nkule field is ideally put into a legume/grain rotation for two to four years, then rested one or two years. Now, many fields of nkule origin are cultivated for six or more years. As in nkomanjila, cassava often ends the cropping sequence, although the wheat/early beans rotation appears to be viable over a long period.

Occasionally an ihombe field will be planted to finger millet a second year or, if well up on the margin, hoed into large ridges for beans and groundnuts. Failure of a second crop of finger millet (or most other crops) may be due to a lack of particular micronutrients in ihombe fields or to alterations of the soil structure. When the soil is cultivated, iron can accumulate in the soil, impeding drainage in subsequent years. A short fallow may reverse this condition.

Andean Agriculture

Between 3,000 and 4,000 years ago, a nomadic, hunting and gathering way of life in the Central Andes was supplanted by a village-based agro-pastoral economy, a system that still prevails despite competition for land between haciendas and peasant communities (Brush 1982). The impact of the complex Andean environment on the human economy has resulted in vertical arrangements of settlements and agricultural systems (see Table 6.2). The pattern of verticality derives from climatic and biotic differences related to altitude and geographical location. The most important cultural adaptation to these environmental constraints has been the subsistence system: crops, animals and agro-pastoral technologies designed to yield an adequate diet with local resources while avoiding soil erosion (Gade 1975).

The evolution of agrarian technology in the Central Andes has produced extensive knowledge about using the Andean environment. This knowledge affected the division of the Andean environment into altitudinally arranged agro-climatic belts, each characterized by specific field and crop rotation practices, terraces and irrigation systems and the selection of many animals, crops and crop varieties (Brush et al. 1981). About 34 different crop species can be found in the area. These include cereals (corn, quinoa, Amaranthus caudetus), legumes (beans, lupine, lima beans), tubers (species of potato, manioc, Arrachocha, etc.), fruits, condiments and vegetables. The main crops are corn, chenopods (Chenopodium quinoa and C. pallidicaule) and potatoes. Individual farmers may cultivate as many as 50 varieties of potatoes in their fields, and up to 100 locally named varieties may be found in a single village. The maintenance of this wide genetic base is adaptive since it reduces the threat of crop loss due to pests and pathogens specific to particular strains of the crop (Brush 1982).

Crop patterns in the agro-climatic belts. The local inhabitants recognize three to seven agro-climatic belts, distinguished according to altitude, moisture, temperature, vegetation, land tenure, crop assemblages and agricultural technology (Table 6.2). There is considerable regional variation in the cultivation patterns of each belt. For example, in the communities of Amaru and Paru-Paru in Cuzco, Peru, three main belts can be distinguished (Gade 1975). Sites in the corn belt have soft slopes, located between 3,400 and 3,600 meters. These sites are irrigated and farmed in three alternative four-year rotations: (1) corn/fava beans/corn/fallow; (2) corn/corn/potato or fallow; and (3) potato and barley/fava beans/corn/corn. The potato/fava/cereals belt is composed of sites with steep slopes, located from 3,600 to 3,800 meters. Potatoes are intercropped with barley, wheat, fava beans and peas. In rainfed areas there are two main four-year rotations: (1) fava beans/wheat/peas/barley and (2) Lupinus mutabilis/barley/fava beans/fallow. In irrigated

Table 6.2 Agroclimatic crop zones of the central Andes (based on Brush 1982)

Zone	Major crops/ animals	Agricultural technology	Land tenure	Focus of production
Pasture above 3800 m.	alpacas llamas sheep cattle	—	communal ownership & communal use	market (esp. wool) and subsistence
Tuber 3000- 4200 m.	potatoes quinoa/canihua barley other native tubers (mashua, ulluca, oca)	hoe foot plow dung as fertilizer	communal ownership w/individual use	subsistence
Cereal 1500- 3000 m.	corn wheat curcurbits beans temperate fruits and vegetables	draft animals some mechan- ization and chemical fertilizers	private ownership and use	subsistence (grains) & market (fruits and vegetables)
Tropical/ fruit 500- 1500 m.	cocoa sugar cane cotton tropical and specialty fruits corn	mainly agro- industrial technology	private ownership and use	market

areas common rotations are: 1) potato/wheat/fava beans/barley and 2) potato or C. quinoa/barley/peas/fallow. The bitter potato/pasture belt is a cold belt located above 3,800 meters. Rainfed rotations in this belt usually include a four- to five-year fallow period, after a four-year sequence of potato/Oxalis tuberosa and Ullucus tuberosus/U. tuberosus and Tropaeolum tuberosum/barley.

Traditional Farming Systems of Mediterranean Chile

The small farmers (campesinos) of mediterranean Chile emphasize diversity to use scarce resources most efficiently. Farming systems are usually either small-scale and intensive or more extensive semi-commercial enterprises.

Small-scale intensive systems. These systems rarely exceed one hectare in size and therefore usually do not provide all the food requirements of the family. All items produced are used on the farm, and other needs are bought with earnings from off-farm work. Campesinos typically produce a great variety of crops and animals, and it is not unusual to find as many as five to 10 tree crops, 10 to 15 annual crops and three to four animal species on a single farm.

These farms often include an arbor of grapes (parron) to provide shade, along with fruit, herbs, medicinal plants and flowers in addition to the tree and annual food crops. The typical animals on these farms are rabbits, free-ranging chickens and ducks and occasionally a few pigs feeding on kitchen waste and crop residue. Intensive annual cropping usually makes use of simple crop patterns (growing annual crops only during the spring and summer), or, more typically, crop sequencing (planting a second crop after the harvest of the first). In both crop patterns, campesinos may practice intercropping. Common intercropping systems include corn/beans, garlic and/or onion mixed with lettuce and cabbage, and corn/potatoes.

Figure 6.3 depicts one very complex system in the central coast range. The land, characterized by a 25 percent slope, was divided into two sections. Half of it was devoted to annual crops and herbs grown in rows running parallel to the hill contour. The other half consisted of a mixed orchard of about 10 species of fruit trees, several varieties of grapes, a few non-crop trees such as pine (Pinus radiata), aromo (Acacia spp.), Datura spp. and a small stand of bamboo and cactus (Opuntia spp.). A living fence of cypress separated the two sections. Chickens and rabbits were raised under the orchard in cages, and their manure, mixed with sawdust, was used to fertilize crops and trees. In addition to the fruit trees, Eucalyptus spp. were planted as a living fence

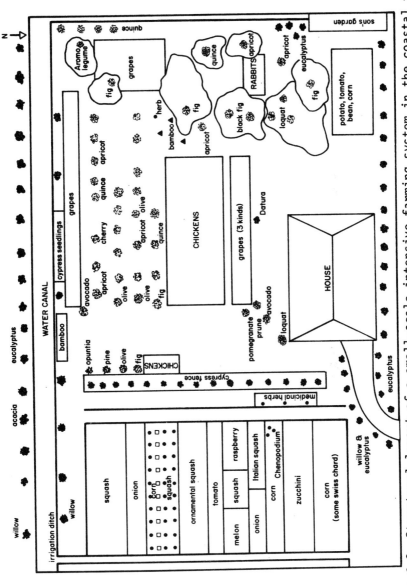

Figure 6.3 Structural layout of a small-scale intensive farming system in the coastal zone of central Chile (Altieri and Farrell 1984).

on the lower boundary and harvested for fuelwood and poles. Additional fuelwood was gathered from the native "espino" (<u>Acacia cavens</u>) growing naturally on the hillside above the property.

Beneath the orchard trees, some herbs were grown for medicinal purposes or to keep chickens healthy, as in the case of Ruda (<u>Ruda tracteosa</u>). According to some campesinos the presence of this plant in the chicken yard prevents infectious poultry diseases. Hinojo (<u>Hinojo officinalis</u>) was allowed to grow freely in the property margins and its cane was later used to construct fences or small huts. Irrigation water was diverted from the canal passing along the upper boundary of the property. The campesinos planted willows (<u>Salix</u> spp.) along the canal to "hold the soil down" and prevent soil sliding. The penetrating root systems, along with the dense canopy cover from the other trees, provided good soil protection on this sloping site.

<u>Extensive semi-commercial systems.</u> Semi-commercial farms range from five to 20 hectares in size. These systems are also diversified, but the crop and animal combinations are designed to increase production to yield a marketable surplus. With more land, the campesino devotes much of it to more extensive activities such as pasture for livestock and grain cultivation. The additional land also affords more space for wood-producing trees. In this way, nearly all of the household requirements are provided for on the farm.

Typically, campesinos grow crops preferred by the local community for commercial purposes. These crops, however, may entail relatively high risks. Therefore they hedge against this risk by growing several less valuable or risky crops, like beans, squash, potato or corn, between rows of high-value fruit trees, like peaches, cherries or apples.

Figure 6.4 shows the design of a 12-hectare farm about 10 kilometers east of Temuco, in south Chile, where the campesino balanced his farm enterprises to provide food, clothing, housing and capital. The farm consisted of an interplanted area of annual crops and fruit trees, a mixed orchard of fruit trees with rows of bee hives between the trees, approximately five hectares of pasture, two to three hectares of wheat and a stand of radiata pine. He harvested 280 kg of honey from 26 beehives per year, obtained 10 to 12 liters of milk per day from three cows, collected 10 to 11 eggs per day from his chickens, and from the wheat, supplied all of his flour for making bread. Pine trees were planted to provide wood. The fast-burning wood was made into charcoal for cooking and heating and was also used in constructing the house and barns. Guano from the animals and crop residues were collected in a compost pile for later use as fertilizer.

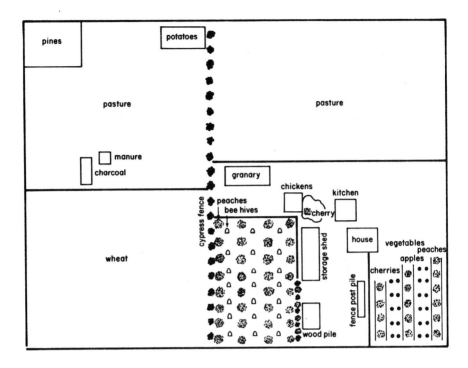

Figure 6.4 Structural layout of a twelve-hectare semi-commercial farming system in southern Chile (Altieri and Farrell 1984).

Raised Field Agriculture

Raised field agriculture is an ancient food production system used extensively by the Aztecs in the Valley of Mexico, but also found in China, Thailand and other areas to exploit the swamplands bordering lakes.

Called "chinampas" in the Aztec region, these "islands" or raised platforms (from 2.5 to 10 meters wide and up to 100 meters long) were usually constructed with mud scraped from the surrounding swamps or shallow lakes. The Aztecs built their platforms up to a height of 0.5 to 0.7 meters above water level and reinforced the sides with posts interwoven with branches and with willow trees planted along the edges (Armillas 1971).

The soil of the platforms is constantly enriched with organic matter produced by the abundant aquatic plants, as well as with sediments and muck from the bottom of the reservoirs. A major source of organic matter today is the water hyacinth (Eichornia crasipes), capable of producing up to 900 kg per hectare of dry matter daily. Supplemented with relatively small amounts of animal manure, the chinampas can be made essentially self sustaining. The animals, such as pigs, chickens and ducks, are kept in small corrals and fed the excess or waste produce from the chinampas. Their manure is incorporated back into the platforms (Gliessman et al. 1981). On the chinampas, farmers concentrate the production of their basic food crops as well as vegetables. This includes the traditional corn/bean/squash polyculture, cassava/corn/bean/peppers/amaranth and fruit trees associated with various cover crops, shrubs or vines. Farmers also encourage the growth of fish in the water courses.

In Asia, raised field agriculture consists of livestock/fowl/fish farming systems. Aquatic vegetation is fed to animals, and in turn their wastes are used as fertilizer for fish ponds and/or vegetable crops grown in platforms adjacent to fish ponds. A common system is pig/fish farming, in which 2,000 to 5,000 kg of fish per hectare are produced every six months. There are about 60 pigs per hectare and fish are stocked at a rate of 25,000 to 30,000 per hectare (Pullin and Shehadeh 1980).

All the traditional agroecosystems described above have proved to be sustainable in their historical and ecological context (Cox and Atkins 1979). Although the systems evolved in very different times and geographical areas, they share structural and functional commonalities (Beets 1982; Marten 1986):

o They combine species and structural diversity in time and pace, through both vertical and horizontal organization of crops.
o They exploit the full range of micro-environments, which differ in soil, water, temperature, altitude, slope and fertility within a field or region.
o They maintain cycles of materials and wastes through effective recycling practices.
o They rely on biological interdependencies that provide some biological pest suppression.
o They rely on local resources plus human and animal energy, using little technology.
o They rely on local varieties of crops and incorporate wild plants and animals. Production is usually for local consumption.
o The level of income is low, so the influence of noneconomic factors on decision-making is substantial.

TRADITIONAL ETHNOBOTANICAL KNOWLEDGE

It is difficult to separate the study of traditional agricultural systems from the study of the cultures that nurture them. It is important to recognize both the complexity and the sophistication of the production systems.

Researchers have documented many complex taxonomic systems used by indigenous people (Berlin et al. 1973). In general, the traditional name of a plant or animal reveals that organism's taxonomic status. Researchers have found good correlation between folk taxa and scientific taxa. In Mexico, the Tzeltals, P'urepechas and Yucatan's Mayans can recognize more than 1200, 900 and 500 plant species respectively (Toledo et al. 1985). Similarly !ko bushwomen in Botswana could identify 206 out of 266 plants collected by researchers (Chambers 1983), and Hanunoo swidden cultivators in the Philippines can distinguish more than 1600 plant species (Conklin 1979).

Polycultures and agroforestry patterns are not developed at random; rather, they are based on a deep understanding of agricultural interactions guided by complex ethnobotanical classification systems. These classification systems have allowed peasants to assign each landscape unit a given productive practice, thus obtaining a diversity of plant products through a multiple-use strategy (Toledo et al. 1985).

In Mexico, for example, Huastec Indians manage a number of agricultural and fallow fields, complex home gardens and forest plots, totalling about 300 plant species. Small areas around the houses commonly average 80 to 125 useful plants, mostly native medicinal plants (Alcorn 1984). Similarly, the traditional Pekarangan in West Java commonly contains about 100 or more plant species. Of these plants about 42 percent provide building materials and fuelwood, 18 percent are fruit trees, 14 percent are vegetables and the remainder constitute ornamentals, medicinal plants, spices and cash crops (Christanty et al. 1985).

TRADITIONAL AGROECOSYSTEMS AND GENETIC RESOURCES

Traditional agroecosystems are genetically diverse, containing populations of variable and adapted land races as well as wild relatives of crops (Harlan 1976). Land race populations consist of mixtures of genetic lines, all of which are reasonably adapted to the region in which they evolved, but which differ in reaction to diseases and insect pests. Some lines are resistant or tolerant to certain races of pathogens and some to other races (Harlan 1976). The resulting genetic diversity confers at least partial resistance to diseases that are specific to particular strains of the crop and allows farmers to exploit different microclimates and derive multiple uses from the genetic variation of a given species.

Andean farmers cultivate as many as 50 potato varieties in their fields and have a four-tiered taxonomic system for classifying potatoes (Brush). Similarly, in Thailand and Indonesia, farmers maintain a diversity of rice varieties that are adapted to a wide range of environmental conditions. Evidence suggests that folk taxonomies become more relevant as areas become more marginal and risky. In Peru, for example, as altitude increases, the percentage of native stock increases steadily. In Southeast Asia, farmers plant modern semi-dwarf rice varieties during the dry season and sow traditional varieties during the monsoon season, thus taking advantage of the productivity of irrigated modern varieties during dry months, and of the stability of native varieties in the wet season when pest outbreaks commonly occur (Grigg 1974). Clawson (1985) described a number of systems in which traditional tropical farmers plant multiple varieties of each crop, providing both intraspecific and interspecific diversity, thus enhancing harvest security.

A number of plants within or around traditional cropping systems are wild relatives of crop plants. Thus, through the practice of nonclean cultivation, farmers have inadvertently increased the gene flow between crops and their relatives (Altieri and Merrick 1987). For example, in Mexico, farmers allow teosinte to remain within or near corn fields so that natural crosses occur when the wind pollinates corn (Wilkes 1977). Through this continual association, fairly stable equilibria have developed among crops, weeds, diseases, cultural practices and human habits (Bartlett 1980). The equilibria are complex, and difficult to modify without upsetting the balance, thus risking loss of genetic resources. For this reason Altieri and Merrick (1987) have supported the concept of in-situ conservation of many land races and wild relatives. They argue that in-situ conservation of native crop diversity is achievable only through preservation of agroecosystems under traditional management, and furthermore, only if this management is guided by the local knowledge of the plants and their requirements (Alcorn 1984).

Many peasants preserve and use naturalized ecosystems (forests, hillsides, lakes, grasslands, streamways, swamps) within or adjacent to their properties. These areas provide valuable food supplements, construction materials, medicines, organic fertilizers, fuels and religious items (Toledo 1980). Although gathering has normally been associated with poverty (Wilken 1969), recent evidence suggests this activity is closely associated with a strong cultural tradition. In addition, vegetation gathering has an economic and ecological basis, as wild plants provide significant input to the subsistence economy, especially when agricultural production is low due to natural calamities or other circumstances (Altieri et al. 1987). In fact, in many areas of semi-arid Africa, peasant and tribal groups maintain their nutritional level even when drought strikes (Grivetti 1979)Gathering is also prominent among shifting cultivators whose fields are widely spaced throughout the forest. Many farmers collect wild plants for the family cooking pot while traveling between fields (Lentz 1986). Gathering is also prevalent in desert biomes. For example, the Pima and Papago Indians of the Sonora Desert supply most of their subsistence needs from more than 15 species of wild and cultivated legumes (Nabhan 1983). In humid, tropical conditions the procurement of resources from the primary and secondary forest is even more impressive. For example, in the Uxpanapa region of Veracruz, Mexico, local peasants exploit about 435 wild plant and animal species, of which 229 are used as food (Toledo et al. 1985).

NOTES

1. Throughout this book the terms traditional, peasant, small-scale and small farning are used synonymously to describe systems that rely on human and animal power and on locally available resources (Wilken 1977).

2. $LER = \frac{P_x}{K_x} + \frac{P_y}{K_y}$ where K_x and K_y are the yields per unit area when the crops are grown in monoculture, and P_x and P_y are the production of the two crop species in a polyculture (Vandermeer 1981).

7

Ecologically Based Agricultural Development Programs

A major technological problem of development projects is that global recommendations frequently prove unsuitable for the conditions of specific peasant farms (de Janvry 1981). The many forms of agriculture found in Third World countries result from variations in local climate, soils, crop types, demographic factors and social organizations, as well as from more direct economic factors such as prices, marketing and availability of capital and credit. What is required is an integrated approach that accounts for these complex interactions. Cropping systems and techniques tailored to specific agroecosystems would result in a more fine-grained agriculture, based on appropriate traditional and improved genetic varieties and local inputs and techniques, with each combination fitting a particular ecological, social and economic niche. Conway (1985) calls this approach agroecosystem analysis and development (AAD).

This approach differs from farming systems research not only in its choice of performance indicators, but also in its level of agroecosystem analysis. AAD deals with all levels of the agroecosystem hierarchy. Studies encompass all interactions between humans and the food-producing resources within both small (field level) and large (regional level) geographic units. A study of the "ecology" of such resources must emphasize the environmental relationships of agriculture, but always within the social, economic and political context. A useful way to examine agroecosystems is to focus on a unifying process such as energy flow, nutrient cycling or population regulation at the community level.

COMBINING TRADITIONAL AND MODERN TECHNOLOGIES

In the Third World, several assistance programs are directed at meeting the subsistence needs of peasants (Hirschman 1984).

Generally planners begin with existing peasant production systems and use modern agricultural science to improve their productivity. The programs have a definite ecological orientation and rely on resource-conserving and yield-sustaining technologies. They emphasize an ecological engineering approach in which the components of agroecosystems—crops, trees, soils and animals-interact in a way that enhances the use of internal resources, recycling of nutrients and organic matter, and trophic relationships among plants, insects, pathogens and weeds that foster biological control (Altieri and Anderson 1986).

Given that in many countries rural populations are becoming increasingly impoverished and have fewer alternatives for generating income (de Janvry 1983), most programs of assistance to peasants are designed as temporary survival strategies to make peasants less dependent on both the market and the state. As programs develop, it is slowly becoming apparent that promoting self-help agricultural strategies can lead to some social change in the absence of large-scale political changes (Hirschman 1984).

Peasant assistance ranges from experimental programs, with impact on a few peasant families, to widely applied programs that affect whole regions. The objective of most of these programs is to build self-help communities that work to improve life at the local level. Most of the agencies providing assistance are staffed by young, middle-class professionals, operating with limited funds provided by private or international foundations. They usually function in isolation from mainstream agricultural colleges and ministries of agriculture.

EXAMPLES OF PROGRAMS

Alley Cropping in Africa

In humid tropical Africa, shifting cultivation and bush fallow systems are still dominant (see Chapter 6). Increasing population and the development of more sedentary agriculture has led to shortened fallows, causing a rapid decline in soil fertility and crop yields. These areas need crop systems based on the use of leguminous plant species to restore soil fertility. In Nigeria, Wilson and Kang (1981) developed a system of alley cropping, an improved fallow system in which selected leguminous shrubs or tree species are planted in association with food crops to accelerate regeneration of soil nutrients, thus shortening the fallow period. In the alley system trees and shrubs provide green manure for the companion crops and pruning materials are used as mulch and shade during the fallow to suppress weeds. Prunings also serve as animal feed, staking material and firewood (Kang et al. 1984). Thus, alley cropping is a multiple-use system.

In these systems food crops are grown in alleys (two to four meters wide) formed by trees or shrubs. Trials where Leucaena leucocephala was intercropped with maize showed substantial increases in crop production. Leaf nitrogen from pruned leaves of Leucaena placed on or incorporated into the soil contributed a significant 23 percent increase in maize yields over the control. The single and double Leucaena rows added on average 100 and 162 kg of nitrogen per hectare respectively to the maize plants. A well-established hedgerow of Leucaena can produce from 15 to 20 tons of fresh prunings (5 to 6.5 tons of dry matter) per hectare with five prunings per year. Three prunings yield over 160 kg of nitrogen, 15 kg of phosphorous, 150 kg of potassium, 40 kg of calcium and 15 kg of magnesium per hectare per year.

Using Manure in Bolivian Andean Agriculture

As in most Andean regions, the staple diet of Bolivia's rural population consists of potatoes and maize. Spanish colonization and recent agrarian reform have radically changed the Inca agricultural system. The use of imported fertilizers in potato growing is becoming more widespread, which means that potatoes are sown more frequently and the land is left fallow for less time. This in turn causes a higher incidence of nematodes and plant diseases, which leads to greater use of pesticides. Average potato production is falling despite a 15 percent annual increase in the use of chemical fertilizer. Due to increases in the cost of fertilizer, potato farmers must produce more than double the amount of potatoes compared with previous years to buy the same quantity of imported fertilizer (Augstburger 1983).

Bolivian peasants have thus become more dependent on agricultural chemicals. Members of the former Proyecto de Agrobiologica de Cochabamba, now called AGRUCO, are attempting to reverse this trend by helping peasants recover their production autonomy. To replace the use of fertilizers and meet the nitrogen requirements of potatoes and cereals, intercropping and rotational systems have been designed that use the native species Lupinus mutabilis. Experiments have revealed that L. mutabilis can fix 200 kg of nitrogen per hectare per year, which becomes partly available to the associated or subsequent potato crop, thus significantly minimizing the need for fertilizers (Augstburger 1983). Intercropped potato/lupine and potato/bean overyielded corresponding potato monocultures, and also substantially reduced the incidence of virus diseases.

In experiments conducted in neutral soils, higher yields were obtained with manure than with chemical fertilizers. In Bolivia, organic manures are deficient in phosphorous. Therefore AGRUCO recommends phosphate rock and bone meal, both of which can be obtained locally and inexpensively, to increase the phosphorous content of organic manures.

The Minka Project in Peru

A group of social scientists under the name Grupo Talpuy, funded by the Fundacion de Tecnologia Andina, have been studying and documenting the traditional farming practices and systems used by peasants in the Peruvian Andes (Brush 1982). The Grupo's main activity consists of rescuing and recording local farming practices like mixed cropping, traditional pest control and fertilization, crop rotations, traditional crop varieties and uses of plants, which are later published in a low-cost magazine called Minka that is circulated throughout the rural areas (Minka 1981). The group also requests information from farmers, extension agents and other people about specific topics.

Each issue of Minka is based on a month-long field survey of a different subject, written very simply and illustrated with drawings and graphics. Subjects have included mixed cropping, Andean crops, local herbal medicine, soil conservation, agricultural tools and low-cost house construction. The magazine promotes the idea that many efficient technologies that originated and are used in local areas can be extended to farmers in outlying areas through the magazine. The objective is to make resources, especially knowledge, widely available. Minka emphasizes the importance of local resources that can be used without specialized knowledge. In this way, farmers can be selective in choosing technologies or practices that have worked for other peasants who share similar levels of capital, land base and natural resources. Follow-up surveys by Grupo Talpuy revealed a great deal of technology adoption and exchange (M. Salas, personal communication).

Clearly peasant-to-peasant technology extension avoids some of the detrimental effects associated with the transfer of foreign technologies (environmental degradation, disruption of subsistence patterns and social relationships). Not all traditional production components are effective or applicable, and Grupo Talpuy understands that modifications and adaptations may be necessary; however, they believe that the foundation of development must remain indigenous.

A Sustainable System for Small Farmers in Chile

In Chile, where subsidized credit has recently been eliminated and technical assistance to farmers privatized, the Centro de Educacion y Tecnologia (CET) is helping peasants achieve year-round food self-sufficiency at low cost. CET's approach has been to establish several half-hectare model farms that can meet most of the food requirements for a family with scarce capital and land. In this system, diversity is the critical factor in using scarce resources efficiently. Thus crops, animals and other farm resources are assembled in mixed and rotational designs to optimize production efficiency, nutrient cycling and crop protection (CET 1983).

The farm consists of diverse combinations of forage and row crops, vegetables, forest and fruit trees and animals. The main components are:
- Vegetables: Spinach, cabbage, tomatoes, lettuce
- Chacras: Corn, beans, potatoes, peas, fava beans
- Cereals: Wheat, oats, barley
- Forage crops: Clover, alfalfa, ryegrass (ballica)
- Fruit trees: Grapes, oranges, peaches, apples
- Forest trees: Black locust, honey locust, willows
- Domestic animals: One milk cow, chickens, pigs, ducks, goats and bees

The family eats the vegetables, fruits and chacras. Forage crops and some chacras serve as food for the animals. Forage crops can also be plowed under as green manure. Fava beans provide the protein in poultry feed. Wheat and oats are used in making bread. All plant residues and manures go into compost. Manure can be applied directly around the base of the fruit trees. Crop residues (such as wheat straw and corn stalks) can be fed to the animals or left on the soil as a stubble mulch.

Non-fruit trees are used for fodder, wood, fuel or construction materials. The tree species black locust (Robinia pseudoacacia) is a nitrogen fixer and also produces pest-resistant wood suitable for fence posts. The foliage of honey locust (Gleditsia triacanthus) and Salix spp. can be used as fodder. Russian wild olive is also a nitrogen fixer and provides a wildlife habitat.

Seedlings are started in a "solar-powered greenhouse," which consists of a big hole in the soil, three by three meters, about one and a half to two meters deep, covered by a sheet of transparent plastic. Most vegetables are produced in heavily composted raised beds. The rest of the vegetables, cereals, legumes and forage plants are produced in a seven-year rotational system described in Figure 7.1. Relatively constant production is achieved by dividing the land into as many small fields of fairly equal productive capacity as there are years in the rotation, which amounts to about six tons per year of useful biomass for 13 different crop species.

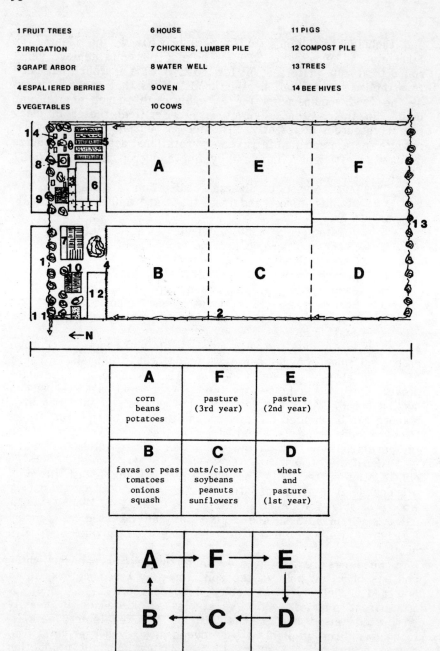

Figure 7.1 Model design of a self-sufficient farming system based on a seven-year rotational scheme adaptable to Mediterranean environments (adapted from CET 1983).

The rotation was designed to produce the maximum variety of basic crops in six plots, taking advantage of the soil-restoring properties of the rotation. In this way each plot receives the following treatments throughout the seven-year period:

Crops	Rotations	Timing
Chacras	Corn/beans/potato	Spring, summer—Year 1
Winter chacras	Peas and fava beans	Fall, winter—Year 2
Vegetables	Tomato, onion, squash, etc.	Spring, summer—Year 2
Supplementary pasture	Oats, clover, ryegrass	Fall, winter—Year 2
Industrial crops	Soybean, peanuts, sunflower	Spring, summer—Year 3
Permanent pasture	Wheat, clover, alfalfa, ryegrass	Fall, winter, spring—Year 4
Permanent pasture	Clover, alfalfa, ryegrass	Summer—Year 5, Fall—Year 7

Crops can be grown in several temporal and spatial designs (such as strip-cropping, intercropping, mixed cropping, cover crops, living mulches) within each plot, thus optimizing the use of limited resources and enhancing the self-sustaining and resource-conserving attributes of the system.

An important consideration in designing the rotation is the stability of the cropping systems, in terms of both soil fertility and pest regulation:

Soil fertility. It is well accepted that rotating grains with leguminous forage crops provides more nitrogen and much higher yields of the subsequent grain crop than are obtained under continuous grain monocropping (see Chapter 11). The output of grain will depend on how efficiently the legumes supply nitrogen. Usually a large tonnage of plant material is disked or plowed into the soil when a nitrogen-fertilizer effect is desired. The tissues incorporated into the soil must be mature. Incorporating mature strains or green manure with high carbon/nitrogen ratios at first results in the "locking up" of soluble soil nitrogen in the cells of the decomposing microorganisms. As a result, additional inputs of nitrogen may be required (Troeh et al. 1980). Studies have shown that legumes such as sweet clover, alfalfa and hairy vetch can produce between 2.3 and 10 tons per hectare of dry matter and fix from 76 to 367 kg of nitrogen per hectare, sufficient for most agronomic and vegetable crops (Palada et al. 1983).

Manure may be applied to the plots in spring or fall. If

applied in the fall it might be immobilized long enough to have a residual effect on the summer crops. Leaving wheat straw in the field might immobilize mineral nitrogen during vegetative growth of beans in the following year, thereby stimulating nitrogen fixation in legumes. The residues decompose during the first few months, although decomposition may be slowed by an inadequate supply of nitrogen (Troeh et al. 1980). Crop residues also provide enough shade to keep the mulched soil cooler than a bare soil, a desirable effect during the summer in an area with a Mediterranean climate.

Pest regulation. The rotational scheme provides nearly continuous plant cover, which aids in controlling annual weeds. In the pasture plots, underseeding wheat with clover helps keep weeds under control after the wheat is harvested. Incorporating legume cover crops in annual crops such as corn, cabbage or tomatoes by overseeding and sod-based rotations has been shown to reduce weeds significantly (Palada et al. 1983). Although these systems may not improve crop yields when compared with clean cultivated crops, they offer great potential for hillside farmers, as they reduce erosion and conserve moisture (see Chapter 10).

Crop rotation also has a profound impact on insect pest populations. For example, more corn rootworms (Diabrotica spp.) are found in a continuous corn monoculture than in corn fields following soybean, clover, alfalfa or other corps. The pest has one generation per year and prefers to oviposit in corn fields. Thus, the environment for a particular pest and its natural enemies might be desirable to improve their synchrony. A compatible winter crop can be responsible for the successful overwintering of large numbers of parasites. Weeds along field margins serve a similar function. Their importance lies in the maintenance of a balance between the pest and its natural enemies during the period the crop is not available. Thus, the annual clean-up of weeds along field edges could destroy overwintering sites of important natural enemies (van den Bosch and Telford 1964).

The presence of alfalfa in the rotational scheme can enhance the abundance and diversity of predators and parasites on the farm. Strip-cutting alfalfa forces predators to move from alfalfa to other crops. Cutting and spreading alfalfa hay containing beneficial insects throughout the farm also increases natural enemy populations (van den Bosch and Telford 1964). The use of cereal residues as straw mulches in the succeeding crops can significantly reduce virus vector white fly populations (Bemisia tabaci) by affecting their ability to attract and alight (Palti 1981).

Infestations of fall armyworm (Spodoptera frugipereda) in corn and of Empoasca spp. (leafhoppers) and Diabrotica spp. (leaf beetles) in beans can be greatly reduced by interplanting both crops (see Chapter 10).

Numerous long-term rotations (three to six years) have been proposed to reduce populations of pathogens in the soil, although short-term sequences can also be effective. Peas, for example, reduce Gaeumannomyces solanacaerum populations that build up during a preceding wheat crop. Incorporating barley straw by rototilling in the topsoil can drastically reduce populations of Verticillium albo-atrum. Incorporating mature legumes or hay as green manure can also affect soil fungal populations and nematodes. Rape, pea or mixed grass legume green manures reduce populations of Gaeumannomyces graminis in wheat by stimulating antagonists (Palti 1981).

Reaching Chilean peasants. Groups of farmers (especially community leaders) coming from local and distant areas live on CET's farms for variable periods of time, learning by participating in the planning, management and evaluation of the organic production systems. After training, farmers are given a packet of the seeds they will need to set up a similar system. They return to their communities to teach their neighbors the new methods and apply the model in their own lands. Follow-up evaluations of the program in rural communities have revealed that many peasants have adopted some or all of the farm design (P. Rodrigo, personal communication). In many instances, peasants have modified the technologies according to their own lore and resources. For example, in southern Chile, a group of peasants did not use compost but instead fertilized their crops with litter from nearby Acacia forests, as is traditional.

The Modular System in the Tabascan lowlands

Various forms of subsistence farming are known to have been employed by the original Indian inhabitants of Tabasco, Mexico, and are thought to have been highly productive (Gliessman et al. 1981). Slash-and-burn agriculture was used for basic grain production (corn, beans), whereas kitchen gardens (huertos familiares), composed primarily of tree crops and their associated understory herbs, shrubs and vines, added great variety to the local diet. Cacao was produced as an understory element in these kitchen gardens and this crop has been expanded considerably with a plantation system that uses legume shade trees.

In recent years the emphasis in agriculture in the Tabascan lowlands has been away from subsistence agriculture and towards commercial farming and stock-raising. Accompanying this shift towards commercial activities has been a gradual abandonment of traditional agricultural practices and varieties. As part of a program to reinstate the diversity and stability of productivity characteristic of the traditional agroecosystems, Gliessman et al. (1981) installed production units, referred to here as modular systems. These systems encourage the application of ecological principles to agriculture through the incorporation of the empirical knowledge present in the region.

Each production unit consists of five to 15 hectares controlled by several families as part of their other agricultural activities. Depending on the social structure of the community, the families may live in the module or may live in a nearby community (ejido) and work in the module during the day. Thus production from each module is either consumed by the families living there, or the products are distributed to the members of the ejido. Any excess in production is available for sale or exchange.

Each production unit has an outermost band of vegetation consisting primarily of second growth species present naturally in the region (Figure 7.2). This band serves simultaneously as a

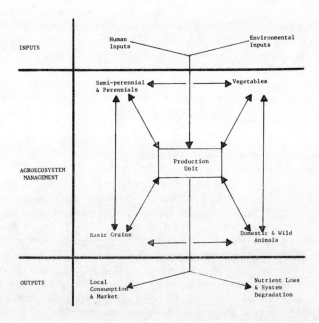

Figure 7.2 Diagrammatic representation of a modular system emphasizing balancing inputs and outputs by various ecological management practices (Gliessman 1982b).

windbreak, a source of natural predators and parasites for biological control and a source of firewood and building materials. At the same time these shelter belts serve as biological reserves or germplasm banks for part of the plants and animals normally present in tropical ecosystems. By selective species enrichment with forest and fruit tree species, it is possible to apply agro-silvicultural management practices, increasing the long-term value of the shelter belt.

The interior part of each modular unit is constructed on the basis of the site's topographic diversity. Where the lowest part of the module can be centrally located, large tanks are constructed to collect the dissolved nutrients and particles of soil and organic matter in the runoff. Fish, ducks and other aquatic animals are produced in the tanks, and the aquatic plants and sediments serve as fertilizer in other parts of the module. Frequently small canals are built radiating out from the central tank to further aid in capturing excessive runoff. To avoid total inundation of the site, a principal canal can be built to eliminate excess water, or in some cases, to add water in times of low rainfall.

Around the central tank or along the edges of the water courses, raised platforms (from 2.5 to 10 meters wide and up to 100 meters long) are constructed, often with the material extracted from the catchment basins, forming a system of "tropical chinampas" for intensive vegetable production. Chinampas are described more fully in Chapter 6.

Agriculture/Aquaculture Systems in Veracruz

In a similar Mexican project, integrated farms were established in the state of Veracruz to help farmers make better use of local resources (Morales 1984). In unique designs based on the chinampas and Asiatic aquaculture systems, vegetable production, animal husbandry and fish production were integrated through the management and recycling of organic matter. The intensive cultivation of corn, beans and squash for local consumption, and of high-value vegetables such as Swiss chard, cabbage, cilantro and chiles, provided abundant plant wastes and cuttings, which were used as cattle and horse feed. All animal wastes were returned as fertilizer for the fields and fish ponds.

Improving Papago Indian Fields

Current efforts by Nabhan (1983) and associates to improve native Americans' arid land agriculture in the United States/Mexico borderlands are based on plants that are adapted to desert conditions and diverse coevolved symbionts (pollinators). These drought-tolerant systems necessarily make use of plants domesticated by the Papago and other Indians over millenia—such as tepary beans (Phaseolus acutifolius), striped cushaw squash (Cucurbita mixta) and devil's-claw (Proboscidea parviflora)—and of plants brought more recently into cultivation that were traditionally harvested from the wild, such as agave, mesquite, jojoba, Mexican oregano and chiltepine. All require less than half the water needed by introduced commercial crops.

IMPLICATIONS FOR THE FUTURE

There is no doubt that, within the range of farming circumstances in the world, and given the present structure of agricultural research and extension, agroecological techniques are more appropriate and adapt and perform better than Green Revolution techniques where natural and socioeconomic resources are scarce. Evidently, the poorer the farmer is, the more relevant low-input approaches are, given that poor farmers have little choice but to use their own resources. Under improved biophysical conditions (good soils, water availabililty) and economic conditions (credit, technical assistance), Green Revolution techniques can be more attractive to farmers, if they can out-yield agroecological strategies in the short term, or provide faster solutions to specific yield-limiting problems (Figure 7.3). This gap would not exist if low-input approaches were supported and subsidized by governments as high-input technologiess have been.

Many traditional farming systems of developing countries contain a wealth of information on efficient crop production under severe resource, biological and socioeconomic constraints. Outstanding features of these systems include their ability to bear risk, and symbiotic temporal and spatial crop mixes that usually result in efficient nutrient recycling and biological pest control. The mechanisms underlying risk aversion, stable crop mixtures, sustained yields, efficient nutrient cycling, pest regulation and other desirable features must be understood to incorporate them into modern crop systems designed to provide increased advantages for small farmers. Recommended systems should enhance the farmers' ability to cope with local and external changes, such as input prices, taxes and government policies (Alcorn 1984). In promoting Green Revolution technologies in the Third World, it is important to remember that farmers lose their autonomy as they become dependent on industry for seeds and other inputs. Thus, rural communities' production systems become governed by distant institutions over which they have little control (Pearce 1975).

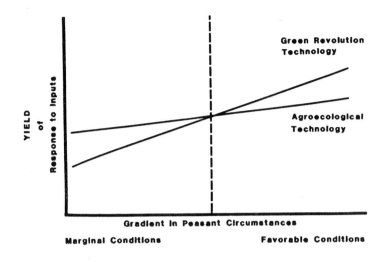

Figure 7.3 The potential performance of Green Revolution technologies (high-input agriculture) and agroecological technologies (low-input agriculture) along a gradient of natural resource and socioeconomic conditions affecting peasant farming systems (Altieri and Anderson 1986).

Data demonstrating that the agroecological projects described in this chapter have improved production, income distribution or employment have been slow to emerge, mainly because the urgencies in the field demand that more time is devoted to action than to research and publication. However, social and biological scientists must collaborate in measuring the success or failure of agroecological projects. More than an analysis of land and labor use and market participation is needed. Researchers must develop a means to measure the achievements of grassroots projects that seek to improve nutrition and well-being by food sharing and communal farming, conserving valuable natural resources and protecting peasants from displacement from their lands and from exploitation as cheap labor.

Gliessman et al. (1981) and Augstburger (1983) assessed the biological relationships and ecological stability of traditional agroecosystems and how farmers improved their total productivity.

CET's farm design provides a creative example of how to assure continuous production of food by effectively organizing the limited space. The follow-up surveys have revealed that peasants adopting the recommended agricultural designs and practices experience fewer food or labor shortages, especially in areas where peasants are organized in activities that reinforce reciprocity and mutual assistance.

In summary, the few examples of grassroots, bottom-up rural development programs described here suggest that development and diffusion of appropriate technologies for peasants must:
- o Start with knowledge of peasant needs as peasants perceive them
- o Use indigenous technologies
- o Be village-based, involving the participation of peasants
- o Emphasize local and indigenous resources (Alcorn 1984).

8

Organic Farming in North America

Organic farming is a production system that sustains agricultural production by avoiding or largely excluding synthetic fertilizers and pesticides. Instead, organic farming relies to the maximum extent feasible on crop rotations, crop residues, animal manures, legumes, green manures, off-farm organic wastes, mechanical cultivation, mineral-bearing rocks and aspects of biological pest control to maintain soil productivity and tilth, to supply plant nutrients and to control insects, weeds and other pests (USDA 1980).

CHARACTERISTICS OF ORGANIC FARMING

The most important difference between organic farming and conventional agriculture is that organic farmers avoid or restrict the use of chemical fertilizers and pesticides in their farming operations, while conventional farmers may use them extensively (Oelhaf 1978). Organic farmers do use modern machinery, recommended crop varieties, certified seed, sound livestock management, recommended soil and water conservation practices and innovative methods of organic waste recycling and residue management.

Research programs on organic farming systems are very limited. Of these, the studies of Lockeretz et al. (1978, 1981), Pimentel et al. 1983 and Oelhaf 1978 on organic farming in the U.S. provide the most comprehensive comparison between conventional and organic agricultural systems. These studies concluded:

o Corn yields were about 10 percent less and soybean yields were about 5 percent less on organic farms than on paired conventional farms. Under highly favorable growing conditions, conventional yields were considerably greater than those on the organic farms. However, under drier conditions, the organic farmers did as well or better than their conventional neighbors. Beyond the third or fourth year after crop rotations became established, organic farm yields began to increase, so that their yields approached those obtained by conventional methods.

o Conventional farms consumed considerably more energy than organic farms largely because they used more petrochemicals. Also, organic farms were considerably more energy-efficient than conventional farms. The researchers found that the energy output/input ratio (or efficiency) for corn production on selected organic farms in 1974 and 1975 was 13 and 20, respectively, while for conventional farms, it was five and seven, respectively. The organic farmers' use of biologically fixed nitrogen and recycled organic wastes significantly reduced energy inputs in agricultural production. However, part of the increased savings from reduced use of fertilizers on organic farms may be offset by their increased use of fuel and machinery to apply manure and cultivate. A recent study comparing the fossil energy, labor and extra land inputs for the production of corn, wheat, potatoes and apples employing organic technologies and conventional technologies suggests the efficiency of energy use in both production systems varies according to the cropping system (Pimentel et al. 1983). The results indicated that organic corn and wheat production were 29 percent to 70 percent more energy-efficient than conventional production. In contrast, organic techniques were 10 percent to 90 percent less energy-efficient than conventional techniques to produce potatoes and apples. Insect pest and disease losses also increased when these crops were grown without pesticides.

o Many organic farms are highly mechanized and use only slightly more labor than conventional farms. Labor requirements averaged 3.3 hectares per acre on organic farms and 3.2 hectares per acre on conventional farms. However, when based on the value of the crop produced, 11 percent more labor was required on the organic farms because the crop output was lower (Lockeretz et al. 1978, 1981). The labor requirements of organic farmers in this study were similar to those of conventional farmers for corn and small grains, but higher for soybeans because more hand weeding was necessary. A number of other studies (Oelhaf 1978, Lockeretz et al. 1975) indicate that organic farms generally require more labor than conventional farms, but exceptions do occur. The labor required to farm organically is a major limitation to the expansion of some organic farms and an important deterrent for conventional farmers who might consider shifting to organic methods.

In many ways, organic farming conserves the natural resources and protects the environment more than conventional farming. Increased public pressure to conserve soil and water and to protect the environment will generate increased interest in organic farming practices in developing countries.

Cropping Systems

Most organic cropping systems include a legume-based rotation with green manure or cover crops (Parr et al. 1983). The usual practice is to follow a heavy green manure crop with a heavily nitrogen-demanding crop (corn, wheat, sorghum). For example, in the corn belt, a typical rotation on organic farms would be three years alfalfa, one year corn (or wheat), one year soybeans, one year corn, one year soybeans, and then back to alfalfa. Legume forage crops provide a source of biologically fixed nitrogen for the organic system. The forage is then fed to animals rather than sold directly, thereby minimizing the flow of nutrients from the farm. Soil productivity is also enhanced by returning animal manure to the land along with crop residues.

Cultural Practices

Most organic farmers use disk or chisel-type implements, which tend to mix the soil rather than invert it. They also practice shallow tillage (six to 10 cm deep), which tends to retain crop residues and manures at or near the soil surface. With shallow tillage, the soil surface is protected by crop residues, promoting water infiltration and storage and reducing soil erosion and nutrient runoff. Successful organic farmers emphasize the importance of timely tillage and planting for weed control and maintenance of good soil tilth. Some organic farmers use delayed planting to control weeds and to increase mineralization of organic matter and release of plant nutrients (Parr et al. 1983).

On organic farms weeds and insects are controlled mainly with nonchemical methods, but with different degrees of effectiveness (USDA 1980). California organic growers combine cultural techniques such as cultivation, crop rotation, smother crops, trap crops, irrigation and solarization (mulching with plastic sheets) with a balanced soil organic matter program and biological control agents to manage pests and diseases (Altieri et al. 1983a).

Weeds are usually a greater problem than insects. Organic weed control methods include crop rotations, tillage, mowing, grazing, competitive crops, intercropping, timely seeding and transplanting, intensive crop spacing and some hand weeding. A description of weed control methods used by organic farmers in California is provided by McLeod and Sweezy (1980). In row crops, weeds are controlled both before and after planting the crop by mechanical and/or manual means. Most farmers are familiar with the life cycles of their crops, and time their cultivation to maximize stress on weeds. Cultivations are kept to a minimum. In the Anderson Valley of northern California one farmer obtained acceptable corn yields by keeping his sweet corn weed-free for only the first four weeks after crop emergence. Several vegetable and herb growers have found the Hydro-Synchron water jet transplanter to be very useful in establishing a crop stand ahead of the weeds (Altieri et al. 1983a).

Mechanical disking and/or mowing are the most common methods used to control weeds in dry-farmed orchards and vineyards. Some growers have obtained encouraging results using articulating mowers. Other growers plant alfalfa or other types of cover as a permanent orchard cover that they mow once or twice a year (Altieri et al. 1983a).

Organic farmers have adequately controlled insects in many field crops using selective crop rotations and natural insect predators. They have experienced great difficulty in controlling insects in vegetable and orchard crops with nonchemical methods. Growers generally favor combinations of organic insecticides and biological methods of pest control (USDA 1980). Beneficial insects commonly released include green lacewings, Trichogramma wasps, predaceous mites, fly parasites, ladybugs, greenhouse whitefly parasite Encarsia formosa, mealybug destroyers, black and red scale parasites, and pink bollworm parasites, which are purchased from local insectaries.

The most common insecticides used are microbial agents, botanical insecticides, oils, soaps and diatomaceous earth. Microbial insecticides such as Bacillus thuringiensis (BT), Nosema locustae and Heliothis nuclear polyhedrosis virus (NPV) products appeal to farmers because of their selectivity. BT is used against leaf- and fruit-feeding lepidopterous larvae such as tomato fruitworm, cabbage looper, diamond back moth, hornworms, codling moth and many others. Nosema disease is used on grasshoppers. Heliothis NPV (nuclear polyhedrosis virus) is used against cotton bollworm and tobacco budworm. Using granulosis virus to control the codling moth has been tested in some organic apple orchards with encouraging results (Falcon et al. 1968). Botanical insecticides include rotenone, pyrethrum, ryania, nicotine sulfate, sabadilla, neem, quassia and wormwood. These are preferred over synthetic chemicals because they are naturally occurring, less toxic and break down relatively quickly in the environment.

Many farmers use dormant and summer oils to smother the eggs of various insects. Mineral oils and rotenone have been used against the codling moth in apple. Recently soap formulations have been tried against codling moth with varied results. Most fruit growers use sex pheromone traps to monitor adult moth pest populations, or contract pest management consultants to do pest scouting and field monitoring.

Fungicides can be used to prevent diseases. These include sulfur, Bordeaux mixture, mined minerals such as copper and calcium carbonate and other formulations made from garlic, horsetail and hydrated lime.

Orchardists find that applying a foliar spray of fish emulsion just before leaf fall aids in leaf decomposition and helps prevent outbreaks of apple scab and other diseases whose spores overwinter on leaves. Scab and powdery mildew are also prevented by applying sulfur/calcium carbonate before and during spring rains.

Most organic farmers also claim that a high proportion of humus in the soil promotes crop resistance to insect pests and plant pathogens because organically grown plants are healthier than commercially fertilized plants, and are therefore more resistant to attack (Bezdicek and Powers 1984). Little research has been conducted to prove this claim, but Culliney and Pimentel (1986) found that late-season population densities of flea-beetles, alate aphids and caterpillars were significantly lower on collards fertilized with sewage sludge and cow manure than on chemically fertilized plants.

Plant Nutrients and Soil Organic Matter

The key to maintaining soil fertility in an organic system is the increased efficiency of nutrient flow from the fixed to the soluble state. Thus organic farmers want to obtain adequate nitrogen and maintain soil organic matter at high levels to ensure maximum soil productivity. The principal source of nitrogen in organic farming systems is atmospheric nitrogen fixed by bacteria associated with legumes. In some cases, off-farm sources of manure or other organic wastes are used. Any nitrogen deficit is decreased further by residual inorganic soil nitrogen, recycling animal manures and crop residues and mineralizing soil organic matter. Materials of low water solubility, such as rock phosphate or greensand (glauconite), are preferred sources of phosphorus and potassium, respectively. Acidulated phosphate sources are sometimes used where rock phosphate is not available.

The USDA (1980) study found that a number of organic farmers apply seaweed and fish emulsion products to the leaves of many field and vegetable crops in the hope that these products provide essential elements for plant growth and protection and increased crop yields.

Most organic farmers believe the amount of organic matter

in the soil is highly correlated with soil productivity and erosion control. Thus, they frequently apply animal manures, and use green manures and cover crops to maintain the soil organic matter. Manure is sometimes composted either in windrows or in static unaerated piles. California apple growers commonly add two tons of compost and one-half ton of limestone to their orchards per acre per year (Altieri et al. 1983a). While returning crop residues to the soil is a common practice on most farms, some farmers actually move residues from one part of the farm to another to increase the level of organic matter where needed.

Because of the great diversity in techniques, climate, soil, management practices, cropping systems and other factors, it is often difficult to compare the nitrogen cycle for organic versus conventional farming techniques. A few general conclusions can be made, however (Power and Doran 1984):

o Organic farming techniques tend to conserve nitrogen in the soil/plant system, often resulting in a buildup of soil organic nitrogen
o Organically managed soils have more soil microorganisms and enhanced levels of potentially mineralizable soil nitrogen
o The net rate of mineralization of nitrogen in organic soils is often slower, resulting at times in mild nitrogen stress during periods of rapid nitrogen uptake
o The presence of organic residues aids in reducing nitrogen losses from the organic agroecosystem
o The effects of organic farming on the nitrogen cycle are most pronounced in the surface soil

Organic Farming and Wildlife

A trend toward organic farming would produce diversity in crop types and smaller fields, benefiting many non-game and game birds. A diversified farming habitat in South Dakota with fencerows, weeds and marshes supported a spring population of 110 pheasant hens per section (2.59 square kilometers) compared with 21 hens per section in a simpler Nebraska habitat comprised of wheat and row crops. Nesting cover, such as winter wheat, pasture and alfalfa (Medicago sativa), comprised 37 percent of the total land area and produced about 63 percent of all pheasant chicks in the simpler habitat of Nebraska. Nesting cover, as oats (Avena sativa), pasture, alfalfa, barley (Hordeum vulgare) and wheat constituted about 48 percent of the land area in South Dakota and produced about 54 percent of all pheasants (Papendick et al. 1986). Enhanced populations of wildlife in agroecosystems can result in increased biological control of certain pests and provide an extra source of income and nutrition if farmers practice selective hunting.

CONSTRAINTS TO ORGANIC FARMING

Although most analysis suggests that organic crop production is more energy-efficient than conventional production, there are several constraints to adopting organic technologies. First, labor productivity usually averages 22 percent to 95 percent less under conventional production. Some researchers may have exaggerated labor costs by including labor input for manure hauling and spreading, but there is no doubt that labor inputs are substantially greater for organic technology. Lockeretz et al. (1981) calculated an increase of 12 percent per unit value of crop produced organically compared with conventional production, and Oelhaf (1978) calculated about a 20 percent greater labor input for organic crops. Pimentel et al. (1983) found that labor productivity was 22 percent to 53 percent less in organic wheat and corn production. Labor productivity for organically grown potatoes and apples was 61 percent to 95 percent less than under conventional production. All these studies used different methods to assess labor inputs, and therefore are not fully comparable. Historically, American farmers have substituted capital for labor and this trend is continuing (Buttel 1980a).

Another constraint is the availability of adequate quantities of organic fertilizer like manure (USDA 1980). For example, only about half of the farms in Iowa keep cattle, which would be a source of manure. This reflects the growing trend toward specialization in U.S. agriculture (Pimentel et al. 1983).

A study by Blobaum (1983) concluded that several obstacles discourage conventional farmers from adopting organic methods. Organic farmers perceive the lack of access to reliable organic farming information as a serious barrier to conversion. Most rely primarily on information from other organic farmers and from such nontraditional sources as books and magazines, representatives of organic fertilizer companies, and workshops and conferences. Organic farmers have a strong interest in research on many problems, including the need for better weed control practices. Most farmers would adopt new practices if more research-substantiated information were available.

Blobaum also found that organic farmers who use special markets are dissatisfied with problems such as small orders, long delays in getting paid, inadequate returns for cleaning and bagging grains, confusing certification standards, difficulty in contacting buyers and the expense of maintaining special on-farm storage areas. Credit discrimination is seen as a potential problem by a sizable number of organic farmers. Problems with credit access, to the extent they exist, appear to involve government farm credit agencies.

Implications of Large-scale Organic Farming in the United States

Langley et al. (1983) used a model to estimate how a complete transformation of U.S. agriculture to organic practices would affect production, supply prices, land use, farm income and export potential. Crop yields and production costs were estimated for 150 producing regions for seven crops under both conventional and organic methods.

Their study concluded that a complete transformation would easily allow the nation to produce enough crops for domestic consumption; however, it would also be necessary to reduce U.S. exports. Net income of the U.S. farm sector would be higher under organic farming because of lower costs of production and higher crop supply prices, but these prices would raise the cost of the domestic food supply. The lower level of production with organic farming methods also implies that the nation's productive reserve would be reduced, which could lead to some shortages in years of relatively poor growing conditions either domestically or abroad. Using net income as a criterion indicates that only the Southeast and southern plains would encounter losses under organic farming.

PART THREE
Alternative Production Systems

9

Polyculture Cropping Systems

Matt Liebman

In many areas of the world, particularly in developing countries, farmers often grow crops in mixtures (polycultures or intercrops) rather than single-species stands (monocultures or sole crops). Until about 10 years ago, the characteristics of polycultures that make them desirable were generally ignored by agricultural researchers. But recently research concerning polycultures has blossomed and some of their benefits are becoming clear.

There is an enormous variety of types of polycultures, reflecting the wide variety of crops and management practices that farmers throughout the world use to meet their requirements for food, fiber, medicine, fuel, building materials, forage and cash. Polycultures may involve mixtures of annual crops with other annuals, annuals with perennials or perennials with perennials. Cereals may be grown in association with legumes, or root crops may be grown in association with fruit trees. Polycultures may be sown in spatial patterns ranging from simple mixtures of two crops in alternate rows to complex assemblies of a dozen or more intermingled species. Component crops of polycultures may be planted at the same date or at different dates (relay cropping); harvests of the components may also be simultaneous or staggered. Different polyculture systems are described in some detail in Papendick et al. (1976), Kass (1978), ICRISAT (1981), Beets (1982), Gomez and Gomez (1984), Steiner (1984) and Francis (1986).

THE PREVALENCE OF POLYCULTURES AROUND THE WORLD

Polyculture cropping systems constitute important parts of the agricultural landscape in many areas of the world. They constitute at least 80 percent of the cultivated area of West Africa (Steiner 1984) and predominate in other areas of Africa as well (Okigbo and Greenland 1976). Much of the production of staple crops in the Latin American tropics occurs in polycultures-more than 40 percent of the cassava, 60 percent of the maize and 80 percent of the beans in that region grow in mixtures with each other or other crops (Francis et al. 1976, Leihner 1983).

Polycultures are very common in areas of Asia where upland rice, sorghum, millet, maize and unirrigated wheat are the staple crops (Aiyer 1949, Harwood and Price 1976, Harwood 1979, Jodha 1981). Lowland rice is generally grown as a monoculture, but in some areas of Southeast Asia farmers build raised beds to produce dryland crops amid strips of flooded rice (Suryatna 1979, Beets 1982).

Although polycultures are prevalent in tropical areas where farms are small and farmers lack capital or credit to purchase synthetic fertilizers, pesticides and field machinery, their use is not restricted to such areas. Polycultures are also used on relatively large, highly mechanized, capital-intensive farms in temperate areas. Examples include the overseeding of forage or green manure legumes into a growing crop of cereal or grain legumes (Hofstetter 1984), strip cropping of soybeans with maize or sunflowers (Radke and Hagstrom 1976) and extensive use of grass/legume mixtures for forage production (Heath et al. 1985).

YIELD ADVANTAGES

One of the major reasons that farmers throughout the world choose to use polycultures is that frequently more yield can be harvested from a given area sown in polyculture than from an equivalent area sown in separate patches of monocultures. This increased land-use efficiency is important in areas of the world where farms are small because of socioeconomic conditions and where crop production is limited to the amount of land that can be cleared, prepared and weeded (usually by hand) in a limited amount of time.

The increased efficiency of a polyculture common in India, sorghum with pigeonpea, is illustrated by data from experiments conducted by Natarajan and Willey (1981). They found that 0.94 hectares of sorghum monoculture and 0.68 hectares of pigeonpea monoculture were needed to produce the same quantities of sorghum and pigeonpea that were harvested from a 1.0-hectare polyculture. The land equivalent ratio (LER) of the polyculture was thus 0.94 + 0.68 = 1.62 (see Mead and Willey 1980 for more information about the LER concept). In this case, the yield of

Table 9.1 Polyculture land use efficiencies, expressed as land equivalent ratios

Polyculture combination	LER	Reference
Millet/peanut	1.26	Reddy and Willey 1981
Maize/bean	1.38	Willey and Osiru 1972
Millet/sorghum	1.53	Andrews 1972
Maize/pigeonpea	.67	Dalal 1974
Barley/fava bean	.85	Martin, Snaydon 1982
Maize/cocoyam/sweet potato	2.08	Unamma et al. 1985
Maize/sweet potato	2.31	Bantilan et al. 1974
Maize/bean/cassava	3.21	Soria et al. 1975

each crop species in the mixture was reduced by competition from the associated crop, but total yield of the polyculture, on a unit area basis, was 62 percent greater than for the monocultures.

LER values reported for a variety of polyculture systems are shown in Table 9.1. It can be seen readily that the land use efficiencies of polycultures can be very large. In the maize/bean/cassava polyculture studied by Soria et al. (1975), 3.21 hectares of monocultures were required to produce the same amount of food that the polyculture produced on 1.0 hectare.

Although farmers often use polycultures without applying fertilizers or pesticides, polyculture yield advantages are not restricted to low-input conditions. High LER values have been reported when large quantities of fertilizers and pesticides have been used (Osiru and Willey 1972, Willey and Osiru 1972, Bantilan et al. 1974, Cordero and McCollum 1979). This is important because it means that farmers with different amounts of resources at their disposal may exploit the increased land-use efficiencies of polycultures.

Some researchers have argued that high land-equivalency values for mixtures of crops with very different maturation times inflate the apparent efficiency of using polycultures, since several short-duration crops might be grown sequentially over the same period of time as a polyculture. These criticisms do not seem to be fully justified, since farmers often need to produce both long-season and short-season crops that can only grow well at certain times of year. Moreover, polyculture yields evaluated in terms of both spatial and temporal efficiency can still show advantages over monocultures (such as beans/cassava, Leihner 1983; maize/cassava, Wade and Sanchez 1984).

In the future, assessments of polyculture performance may include several different criteria, including calories and protein production per hectare per day (Wade and Sanchez 1984). These indices more closely approximate the criteria used by farmers for choosing cropping systems best able to provide a diverse, nutritious diet and marketable materials.

Consideration of the economics of different crop systems has shown that net economic returns from polycultures can be higher than from monocultures grown over equivalent areas. Norman (1977) studied cropping systems in northern Nigeria and found that when the cost of labor was included in his analyses, profitability was 42 percent to 149 percent higher for polycultures than for monocultures. Leihner (1983) found that in Colombia more labor was required for cassava/bean polycultures than for sole-cropped cassava, but that net incomes from the polycultures were higher. Profitability of crop systems can change dramatically from year to year, however. Sanders and Johnson (1982) noted that in one year monoculture bean systems were more profitable than maize/bean polycultures, but in the following year, when prices of the two crops changed, the relative profitabilities of the two systems were reversed.

Yield Stability

In farm systems where subsistence is the primary objective, reducing the risk of total crop failure appears to be at least as important as increasing potential nutritional and cash returns (Lynam et al. 1986). Yield variability of cereal/legume polycultures can be lower than for monocultures of the components, as found for six different mixtures (Papadakis 1941) and for sorghum/pigeonpea mixtures (Rao and Willey 1980). Thus, the likelihood of having nothing to eat or sell is apparently less when crop mixtures are used. Indeed, Trenbath (1983) has shown that for a given land area, the probability of a family failing to produce enough calories for subsistence is lower when the area is sown to a sorghum/pigeonpea polyculture than when it is sown to monocultures of the same components. Francis and Sanders (1978), working with maize and beans, and Rao and Willey (1980), working with sorghum and pigeonpea, found that the probability of exceeding a specified "disaster income level" was greater for polycultures than monocultures.

Trenbath (1976) suggested that yield compensation may occur between polyculture component crops, so that failure of one component due to drought, pests or other factors might be offset by increased yield by the other component(s). Kass (1978) cited a study by Gliemeroth (1950) illustrating this principle. When oat stands were reduced due to an attack by wireworms, the yield of peas sown with oats was greater than the reduction in the oat yield; oat yield was reduced by one-half while pea yield increased fourfold. Other data demonstrating this type of compensation phenomenon are fairly scanty (Harwood 1979). Much more research is needed before increased yield stability can be assumed as a general characteristic of polycultures; in cases where stability does increase, more research is needed to understand the causal mechanism.

Resource Use

In contrast with the state of knowledge on polyculture yield stability, significant progress has been made in understanding the mechanisms that contribute to polyculture yield advantages over monocultures. As researchers direct attention toward patterns of crop growth and resource use in polycultures and monocultures, it is becoming clear that yield advantages by polycultures are often correlated with use of a greater proportion of available light, water and nutrient resources.

When total densities are higher in polycultures than in monocultures, polycultures can intercept more light early in the growing season. This phenomenon has been observed in mixtures of maize with mungbean, peanut or sweet potato (Bantilan et al. 1974), and sorghum with cowpea, mungbean, peanut or soybean (Abraham and Singh 1984). Polycultures composed of crops with nonsynchronous patterns of canopy development and different maturation times (such as sorghum/pigeonpea mixtures, as studied by Natarajan and Willey 1980) can display a greater amount of leaf area over the course of the growing season and intercept more light energy than monocultures.

The greater amount of canopy cover produced by polycultures can decrease penetration of sunlight to the ground surface so that a greater proportion of available soil water is channeled through the crops as transpiration, rather than being lost as evaporation from the soil; Reddy and Willey (1981) observed this with millet/peanut mixtures. Increased canopy coverage by polycultures can also increase penetration of rainfall into the soil and decrease soil erosion by lessening the impact of rain on the soil surface, such as with maize/cassava mixtures (Lal 1980).

Polycultures composed of species with spatially complementary rooting patterns can exploit a greater volume of soil and have more access to relatively immobile nutrients like phosphorous (O'Brien et al. 1967, Whittington and O'Brien 1968). Polycultures composed of species that have temporally complementary patterns of root growth and nutrient uptake can capture more nutrients than monocultures if these nutrients are being made available continuously through mineralization.

Natarajan and Willey (1980) observed this phenomenon with sorghum/pigeonpea mixtures, as did Reddy and Willey (1981) with millet/peanut mixtures. Annual crops grown in association with trees may benefit when leaves of the deeper-rooted perennials fall and decompose, releasing nutrients tapped deeper in the soil profile, as with millet planted beneath Acacia albida trees (Charreau and Vidal 1965).

If one of the component species in a polyculture is a legume bearing nitrogen-fixing bacteria on its roots, atmospheric nitrogen may be transferred to associated non-legumes and increase their yield considerably. Agboola and Fayemi (1972) observed this phenomenon with maize/mungbean mixtures, as did Kapoor and Ramakrishnan (1975) with wheat/Trigonella polycerata mixtures and Eaglesham et al. (1981) with maize/cowpea mixtures. Even when nitrogen is not directly transferred between legumes and non-legumes, fixation of atmospheric nitrogen by the legume to satisfy its own requirements may spare soil nitrogen supplies for use by the associated non-legume(s) (deWit et al. 1966, Martin and Snaydon 1982). Although polyculture yield advantages are more common under conditions of low soil nitrogen availability, application of nitrogen fertilizer does not necessarily eliminate polyculture yield advantages. Large polyculture yield advantages have been obtained when nitrogen fertilizer has been applied at rates (130 kg nitrogen per hectare) considered adequate to satisfy fully polyculture demand (Willey and Osiru 1972, Osiru and Willey 1972).

In some cases, growing crop plants in polycultures can cause the plants to direct proportionately more of the materials they acquire through photosynthesis and root uptake toward the harvested portions of their bodies. For instance, Natarajan and Willey (1981) observed that when pigeonpea was grown alone, its seeds constituted 19 percent of its total above-ground plant weight. However, in polyculture with sorghum, pigeonpea allocated 32 percent of its total above-ground weight to seeds, a 68 percent increase. The increased allocation of carbon and nutrients to seeds meant that seed yield of the intercropped pigeonpea plants was quite high, even when their overall size was greatly reduced by their association with sorghum.

Of particular interest are the results of Natarajan and Willey (1986) from experiments with monocultures and polycultures of sorghum, millet and peanut. The researchers found that increases in allocation ratios for these crops that occurred when they grew in polycultures were most marked under drought conditions. The polycultures were most advantageous for seed yield when water availability most severely affected overall plant productivity.

EFFECTS OF POLYCULTURES ON INSECT PESTS, DISEASES, NEMATODES AND WEEDS

Insect Pests

Scientific documentation shows that insect pests are frequently less abundant in polycultures than monocultures (see Chapter 13). Risch et al. (1983) reviewed 150 published field studies and found that 53 percent of the pest species in the survey were less abundant in polycultures, 18 percent were more abundant in polycultures, 9 percent showed no difference and 20 percent showed a variable response.

Diseases

Little research has yet been done on the influence of polycultures on plant diseases (Sumner et al. 1981). In some cases the incidence of disease is higher for crops grown in polycultures than monocultures; in others the reverse situation occurs (see also Chapter 13). For instance, in experiments conducted in Costa Rica, Moreno (1975) found that, compared with a monoculture of cassava, the severity of cassava mildew was greater when cassava grew with maize but lower when cassava grew with beans or sweet potato. Moreno (1979) also found that the severity of angular leafspot on beans was greater in association with maize but lower in association with cassava or sweet potato, compared with a monoculture of beans.

Researchers are just beginning to understand the underlying mechanisms that affect diseases in different crop systems. The following aspects of polycultures may be important for improving plant health:

1. Susceptible plant species can be planted at lower densities in polycultures than monocultures since the space between them can be occupied by resistant plant species that are valuable to the farmer. Decreased density of susceptible plants can decrease the spread of diseases by reducing the amount of tissue that is infected and subsequently serves as a new source of inoculum. For some diseases, increasing the distance between susceptible plants by reducing their density can also reduce the spread of inoculum. This was noted for monocultures and mixtures of barley and wheat exposed to barley mildew (Burdon and Whitbread 1979).

2. Resistant plants interspersed among suceptible plants can intercept disease inoculum spread by wind or water and prevent it from infecting the susceptible plants (the "flypaper effect"). Moreno (1979) suggested this as the mechanism for the decreased incidence of <u>Ascochyta</u> <u>phaseolorum</u> on cowpea when this crop

was sown in association with maize.

3. The microclimate of polycultures may be less favorable for disease development. Reduced severity of several pea diseases has been observed when pea vines climb up associated cereals, rather than lying matted on the ground (Johnston et al. 1978). Intercropping the peas with the cereals improves air circulation and reduces humidity. In other crop mixtures, denser canopy coverage may increase humidity and reduce light penetration, favoring certain fungal and bacterial diseases (Palti 1981). The latter effect may require the use of spatial arrangements in polycultures that promote a more open canopy configuration.

4. Excretions from or microbes on the roots of one crop species may affect soil disease organisms that attack roots of an associated crop species. This appears to be the mechanism responsible for the decreased incidence of Fusarium udum wilt of pigeonpea when this crop grew in polycultures with sorghum (ICRISAT 1984).

Nematodes

Very little research has focused on the effects of polycultures on pest nematodes (see Chapter 9). However, it has been well established that nematodes prefer particular crop species (Palti 1981) and that certain plants, such as marigolds (Tagetes), excrete substances that are toxic to nematodes (Cook and Baker 1983). These effects suggest it may be possible to decoy, trap or kill nematodes by interplanting certain crop species amid other crops that require protection. Visser and Vythilingum (1959) reported that marigolds growing between tea bushes reduced populations of nematodes in the soil and in tea roots. When the legume cover crop Crotalaria spectabilis was sown in peach orchards, nematodes attacked the legume rather than the tree crop, increasing fruit yields (Cook and Baker 1983).

Weeds

Weed control is one of the most labor-intensive aspects of tropical agriculture and one of the most chemical-intensive aspects of temperate agriculture. Very little research has been conducted concerning the effects of polycultures on weeds, but there are some data showing that using polycultures may provide some weed control advantages over monocultures. Polycultures appear to be most advantageous for weed control when the farmer is most interested in the yield of a main crop and intersows additional species for weed control, erosion control, improved soil fertility and a small amount of additional crop yield. When the farmer is interested in yield of all the component species of a polyculture, weed suppression by the polyculture is frequently intermediate between that obtained from the respective monocultures.

Akobondu (1980) reported that in terms of crop yields and weed suppression, smother crops of egusi melon and sweet potato could replace three hand weedings when they were sown together into sole-cropped yam, sole-cropped maize and polyculture combinations of yams, maize and cassava. The vining smother crops in these Nigerian experiments not only served as a labor-saving means of weed control, but also provided erosion control through increased soil coverage.

In experiments conducted in India, Shetty and Rao (1981) found that adding smother crops of cowpea or mungbean to main crops of sorghum or pigeonpea resulted in less early-season weed growth and decreased the number of hand weedings necessary for high crop yields from two to one. The smother crops had no effect on yield of the main crop species and provided additional yield themselves.

Abraham and Singh (1984) measured the crop yield and weed suppression effects of adding cowpea, peanut, soybean or mungbean to sorghum. Intersowing any of the four annual legumes increased yield and nitrogen content of sorghum and depressed weed growth below levels in sole-cropped sorghum. Forage or seed yield of the legumes was an additional benefit. Similar results were obtained by Tripathi and Singh (1983) when they added soybeans to maize.

In temperate climates, overseeding green manure legumes into cereal and grain legume crops can provide increased weed control for the main crops and furnish ground cover for erosion control during the fall and winter. When an oversown legume is plowed down the following spring, the atmospheric nitrogen it has fixed can be used by succeeding crops. Harwood (1984) reported that in Pennsylvania, oversowing red clover or hairy vetch into maize or soybeans (planted 35 days earlier and cultivated once) had no effect on yields of the concurrent grain crops, greatly reduced weed growth and created nearly complete soil cover. The oversown red clover and vetch fixed an estimated 146 kg of nitrogen per hectare and 367 kg of nitrogen per hectare, respectively. A nitrogen fertility experiment conducted at an adjacent site showed that legume cover crops provided enough nitrogen for high yields of subsequent maize crops (Palada et al. 1983).

The factors that affect the success of weed control in polycultures are poorly understood. Moody and Shetty (1981) suggested that increased suppression of weeds in polycultures as compared with monocultures was largely the result of greater overall crop density in polycultures. Results of experiments conducted in India with weedy sorghum/pigeonpea polycultures support this hypothesis: when density of the polyculture equaled that of the monocultures, weed growth in the polyculture was greater than in sole-cropped sorghum and less than in sole-cropped pigeonpea. But when the polyculture density exceeded that of the monocultures, weed growth in the polyculture was lower than in both monocultures (Shetty and Rao 1981). In contrast, results of experiments conducted in Brazil with weedy sunflower/bean, maize/bean, and maize/sunflower polycultures showed that polycultures may provide superior weed control without an increase in total crop density above monoculture levels. Polyculture and monoculture densities were equal but polycultures always contained less weed biomass than did monocultures of the component crops (Fleck et al. 1984). Thus, the relative importance of increasing crop density and diversity may vary considerably with different crop systems and, presumably, different weed species.

Superior weed suppression by polycultures is often attributed to denser canopies that intercept light that otherwise would reach weeds. Evidence concerning the relationship between light interception and weed suppression is not consistent, however. Bantilan et al. (1974) found that, early in the growing season, maize/peanut, maize/mungbean and maize/sweet potato polycultures all intercepted more light than did the respective monocultures. But only the maize/mungbean polyculture suppressed weed growth more than did both components in monoculture; weed growth in the maize/sweet potato and maize/peanut polycultures was intermediate between that found in the respective monocultures. These data indicate that increased preemption of light may be important for weed suppression by some polyculture systems but not others.

The experiments of Bantilan et al. (1974) also provide an example of the complexity of polyculture/weed relations that is possible when soil fertility varies. The researchers observed that as nitrogen fertilizer was applied at increasing rates, weed weight increased 125 percent and 67 percent in maize/peanut and maize/sweet potato polycultures, respectively, but decreased 59 percent in a maize/mungbean polyculture. Since the weed species involved in the polyculture experiments had been shown to respond very positively to nitrogen in a related group of experiments (Bantilan et al. 1974), it can be concluded that nitrogen increased competitive suppression of weeds by the maize/mungbean polyculture, and that it decreased or had no effect on competition from the other two intercrop systems.

Much more information is needed concerning the effects of different crop cultivars, weed species, soil fertility and moisture conditions, crop spatial arrangements and densities, and other factors on polyculture/weed interactions. Some of the existing information can be found in Moody and Shetty (1981), Altieri and Liebman (1986) and Liebman 1988 (in press).

FUTURE DIRECTIONS

Despite the lack of research concerning many aspects of polyculture cropping systems, it is clear from the preceding discussion that in many situations polycultures can provide farmers with some important benefits. The prevalence of polycultures in developing countries suggests that many farmers there are well aware of these benefits. It seems extremely counterproductive to try to convince these farmers to abandon the use of polycultures where and when benefits can be obtained.

Researchers working in developing countries should develop crop varieties and management practices that are compatible with and improve the performance of polyculture systems (Francis et al. 1976; Krantz 1981). One example of an appropriate technology for polycultures is the design and production of low-cost, animal-drawn planters and cultivators specifically for crop mixtures (Anderson 1981). Pest control and soil fertility aspects of polyculture systems deserve much more attention in developing countries where access to synthetic pesticides and fertilizers is limited by socioeconomic conditions and considerations of human and environmental health.

The role of polycultures in the agriculture of developed countries will probably expand as there is increased perception of the economic and environmental costs of heavy reliance on agricultural chemicals (Horwith 1985). Although agriculture in these countries is extensively mechanized, polyculture systems can be compatible with mechanization (such as green manure legumes oversown into grains; soybeans relay-cropped with wheat; forage mixtures of grasses and legumes; cover crops for orchard floors). As in the developing countries, crop varieties and management practices are needed that will enhance the benefits of existing polyculture systems. Increased attention to the design of machines for other types of crop mixtures might allow the potential biological benefits of these systems to reach farmers in a practical way.

NOTES

1. Department of Plant and Soil Sciences, University of Maine, Orono.

10

Cover Cropping and Mulching

Cover cropping is the practice of growing pure or mixed stands of annual or perennial herbaceous plants to cover the soil of croplands for part or all of the year. The plants may be incorporated into the soil by tillage, as in seasonal cover cropping, or they may be retained for one or several seasons. When plants are incorporated into the soil by tillage, the organic matter added to the soil is called green manure.

BENEFITS OF COVER CROPPING

The possible benefits of cover cropping in orchards and vineyards include (Finch and Sharp 1976, Haynes 1980):
 o Improves soil structure and water penetration because adding organic matter and roots increases soil aeration and the percentage of water-stable aggregates. Tillage requirements and equipment travel are decreased, thereby reducing soil compaction and tillage pan. Vegetative cover can better support machinery during wet periods. The cover crop intercepts water drops, reducing their force and preventing crust formation.
 o Prevents soil erosion by spreading and slowing the movement of surface water, reducing runoff and holding the soil in place with root systems.
 o Improves soil fertility by adding organic material to the soil during decomposition and by making nutrients in the soil more available through nitrogen fixation.
 o Controls dust by holding the soil in place with root systems.
 o Aids in controlling insects by harboring beneficial insect predators and parasites.
 o Modifies the microclimate and temperature by reducing

reflection of sunlight and heat and increasing humidity in the summer.
  o Minimizes competition between the main crop and noxious weeds.
  o Reduces soil temperatures.

The value of cover crops in maintaining soil fertility in orchards depends partly on the production of reasonably heavy tonnages of organic matter. Purple vetch can produce 20 tons of green manure per acre, whereas other legumes produce from 12 to 13 tons per acre. Purple vetch and sweet clover can produce net gains of nitrogen of up to 150 pounds per acre per year.

Four different cover management systems were widely tested in Malaysia's rubber tree (Hevea) plantations—a mixture of creeping legumes (Calopogonium muconoides, Centrosema pubescens and Pueraria phaseoloides), grasses (mostly Axonopus compressus with Paspalum conjugatum), a pure crop of Mikania cordata, and a naturally regenerating system representing the normal colonization process on cleared land. Of the four systems, legumes initially had the fastest rate of growth, and generally contained more nutrients than the other covers tested. The greater nutrient return to the soil from a leguminous cover was reflected in higher levels of these nutrients in rubber leaves. Coupled with improved soil physical properties, this led to an increased rate of growth of the rubber tree. Nitrogen fixation under legumes grown in association with rubber averaged 150 kilograms per hectare per year over a five-year period. Maximum rates of nitrogen fixation were about 200 kilograms per hectare per year.

Two hypotheses may explain these effects: first, that legumes recycle nutrients at or near the soil surface until they can be efficiently used by Hevea, and second, that legumes, by processes not fully understood, cause increased proliferation of Hevea roots, which facilitates nutrient uptake (Broughton 1977).

Plants that are useful under some conditions may be a liability under others. Cover crops used in orchards and vineyards may compete with trees or vines for water and nutrients, and certain weeds may proliferate, reducing the cover crop stand substantially. In areas where it is impractical to grow legumes, it may be advisable to turn to mustards, malva and rape. These plants contain large percentages of nitrogen and quickly decompose if turned under before reaching maturity. Mustards grow very rapidly and can choke out other undesirable weeds. Cover crop plant residues may also interfere with harvesting of fruits and nuts.

Soviet researchers found that the effectiveness of the

parasitic wasp Aphytis proclia in controlling the San Jose scale improved as a result of planting a Phacelia tanacetifolia cover crop in the orchards. Three successive plantings of the Phacelia cover crop increased parasitization of scales from five percent in clean cultivated orchards to 75 percent where honey plants were grown and in full bloom (Altieri and Whitcomb 1979).

In northern California, manipulation of ground cover vegetation in apple orchards and vineyards had a substantial impact on the abundance of soil-dwelling and foliage-inhabiting arthropods. Systems with cover crops were generally characterized by lower densities of phytophagous insects, less fruit damage caused by insects on the trees, larger populations and more species of natural enemies and increased predation of artificial prey. Cover crops that remained in full bloom throughout the season, that produced more biomass and supported higher numbers of alternative prey, seemed to harbor the largest complex of predators and parasites. Apparently, cover crop manipulation can directly affect colonization of insect pests that discriminate among trees with and without cover beneath, and can also help retain populations of natural enemies that inhabit soil and foliage by providing alternative food and habitat. The design of proper cover crop/orchard mixtures can enhance biological control of specific pests in existing orchards and vineyards (Altieri and Schmidt 1985).

TYPES OF COVER CROP MANAGEMENT

The drawbacks of cover crop systems can be reduced or eliminated with careful management and agronomic practices. Limitations are small compared with the alternatives. The most common cover crop management systems are (Finch and Sharp 1976):

Nontillage Systems

Under a nontillage management system the cover crop is mowed rather than disked into the ground. Nontillage reduces soil compaction and soil erosion and improves water infiltration.

A nontillage system can be started in an existing or new orchard. An existing orchard should be properly prepared soon after harvest. It is particularly important to do a good job of leveling and grading, since the soil will not be reworked. For initial planting of a cover crop, Table 10.1 provides recommended seedage per acre and methods of sowing for different species proper for California orchards and vineyards.

Frequent clipping. In this system the cover crop is clipped

four to seven times, beginning in early spring. This system is used with drag-hose operations and with sprinkler, border, furrow and drip irrigation systems. Frequent mowing eliminates the use of many deep-rooted, reseeding annual and perennial plants. Low-growing, reseeding annuals or perennials do best under this type of management.

Infrequent clipping. In this system the cover crop is infrequently clipped, usually in early spring for frost protection and in late spring for residue control. It is not well adapted to drag-hose irrigation. This system permits the use of deep-rooted, reseeding annual or perennial plants. If reseeding annuals are used, spring mowing must be timed to allow a crop of seed to mature for the next year's stand. Through close and careful management the danger from frost or residue buildup can be minimized.

Tillage Systems

Under a tillage system, the cover crop is disked into the soil after the seeds have matured. The optimal timing for tillage is given for various species in Table 10.1.

Annually fall-seeded cover crop. In this system fall-seeded cover crops are disked into the soil in early spring, followed by either summer fallow until fall, or volunteer summer annuals. Early tillage is used to turn under the green manure crop and reduce danger of frost damage. This system can be used with all types of irrigation in most orchards and vineyards. Frequent tillage is a disadvantage of this system. Only short-season annual plants can be used, and the soil is exposed for much of the year.

Reseeding winter annual cover crop. In this system reseeding winter annuals are disked down in late spring, followed by either summer fallow until fall, or volunteer summer annuals, which are mowed, then disked down in the fall. The cover crop can be clipped until late spring to control vegetation height. Mowing must be timed to allow the reseeding annuals to produce a mature seed crop before disking them down. Many reseeding, deep-rooted annuals are ideal for this system.

No winter cover. In this system winter cover is eliminated by cultivation or chemical control. This is followed by either volunteer summer annuals, annually summer-seeded annuals or reseeding summer annuals. The summer cover is used from mid-spring until frost. This system works well with border, furrow or sprinkler irrigation. It is most frequently used in table grape vineyards and has possible use in citrus.

In some citrus-growing areas, particularly in Florida, cover crops are useful in summer because it is the season of greatest

Table 10.1 Partial list of species and some management characteristics of cover crop plants recommended for California orchards and vineyards (after Finch and Sharp, 1976)

Cover crop species[1]	Planting density (seed lb/acre)	Mgt. system	Disked/Mowed		Special Features
Barley (Hordeum vulgare)	90	Tillage	Spring	---	Fast growing in winter
Cereal rye (Secale cereale)	60	Tillage	Spring	---	" "
Annual ryegrass (Lolium multiflora)	9	Tillage	Spring	---	Late maturing winter annual
Purple vetch (Vicia atropurpurea)	52	Tillage	Spring	---	N fixer
Blando brome (Bromus mallis)	6	No till	---	Spring[2]	Good reseeding ability
Cucamonga brome (Bromus carinatus)	12	No till	---	Spring[2]	Matures in April
Wimmera 62 ryegrass (Lolium rigidum)	9	No till	---	Spring[2]	Well adapted to lowlands
Annual bluegrass (Poa annua)	5	No till	---	Frequent mowing	Matures in early spring
Lana vetch (Vicia dasycarpa)	15	No till	---	Infrequent mowing	Reseeds well
Rose clover (Trifolium hirtum)	9	No till	---	Spring[2]	Early maturing, poor competitor
Crimson clover (T. incarnatum)	9	No till	---	Frequent mowing	Adapted to acid soils
Bur clover (Medicago hispida)	9	No till	---	Frequent mowing	Reseeds well
Black medic (Medicago sp.)	6	No till	---	Frequent mowing	Adapted to alkaline soils

1. All cover crops are planted in fall in California.
2. Proper for frequent mowing but must be allowed a regrowth of 3-4 weeks prior to seed maturity.

rainfall. In other areas, such as California, the heavy rains come in winter, which may be the only season when it is practical to grow a cover crop. In the large irrigated areas of California, the water supply is insufficient to grow a cover crop in summer and also provide for the trees' moisture requirements. A cover crop of 10 tons per acre may require 12 inches or more of water per acre.

Disposal of Cover Crop

For a cover crop to be beneficial, it must decay in the orchard or vineyard. To promote decomposition, the material must be incorporated with damp soil. Therefore, it might be advisable to turn under a cover crop deeper than that provided by the shallow summer cultivation. Care should be exercised, however, to make sure that plowing and disking is not so deep as to cut many tree roots. All orchard disks should be equipped with rollers to prevent excessive penetration. It may sometimes be desirable to break down a large cover crop with a drag or disk before working the crop into the soil. This procedure makes plowing or final disking easier, and lessens the loss of water by transpiration- a result to be desired if the soil is drying out faster than the cover crop can be turned under.

Cover Crop Plants

A good cover crop plant maintains or improves soil conditions while it satisfies the soil, site and management requirements of a particular orchard or vineyard. The wide variety of management systems in orchards and vineyards creates demand for a diversity of cover crops. Grasses have fibrous root systems that make them particularly useful in building soil structure, providing erosion control and improving water penetration. Legumes are not as effective as grasses in improving water penetration but they do contribute nitrogen to the soil and their residues break down more rapidly. Plants useful as cover crops can be classed as annually seeded winter-growing grasses and legumes, reseeding winter annual grasses and legumes, summer annuals, perennial grasses and legumes, and other cover crop plants.

LIVING MULCHES

Using legume cover crops in year-round cropping systems and rotations offers a great potential for sustained crop production and self-sufficiency in soil nutrients. Legume cover crops used in association with annual crops are generally called living mulches. Most research on these systems has been conducted on corn, soybeans and vegetable crops in the form of legume overseeding, sod rotation and sod interplanting (Miller and Bell 1982).

Legume species commonly used as living mulches include short white clover, hairy vetch and red clover. Growth characteristics of representative legumes usually used as living mulches are presented in Table 10.2. Except for alfalfa, most legume species are annuals or biennials. Adaptations range from semi-temperate for hairy vetch and crimson clover to temperate for alfalfa, winter pea and sweet clover. Dry matter production ranges from 2.3 tons per hectare for sweet clover to 10 tons per hectare for alfalfa and hairy vetch. Based on tissue nitrogen content and dry matter production, these legumes fix from 76 to 367 kg of nitrogen per hectare. This is sufficient to meet most

Table 10.2 Growth characteristics of legume species used as cover crops (after Palada et al. 1983)

Common name	Scientific name	Growth habit[a]	Adaptation	Dry matter (t/ha)	Total N (kg/ha)
Alfafa	Medicago sativa	P	temperate	10.0	170
Crimson clover	Trifolium incarnatum	A	semi-temperate	7.9	179
Hairy vetch	Vicia villosa	A	semi-temperate	10.2	376
Medium red clover	T. pratense	B,P	semi-temperate	5.2	146
Short white clover	T. repens	B	semi-temperate	5.2	182
Yellow sweet clover	Melilotus officinalis	B	temperate	2.3	76
Winter pea	Pisum sativum subsp. arvense	A	temperate	6.0	213

a) A = annual, B = biennial, P = perennial

of the nitrogen requirement of agronomic and vegetable crops (Palada et al. 1983).

Most legume cover crops do not tolerate acid or dry soil but do tolerate shade and field traffic, which are ideal characteristics for intercropping. Resistance to severe winter frost is important if the legumes are to be grown for soil nitrogen. Winter survival and spring regrowth seem to be fair with selected species.

CROPPING SYSTEMS WITH LEGUME COVER CROPS

Legume cover crops can be incorporated into year-round cropping systems by overseeding (also termed interseeding), legume sod-based rotations, sod strip intercropping or vegetable living-mulch systems (Palada et al. 1983).

Legume Overseeding

Overseeding legume cover crops into small grains in the spring has been a standard farming practice for several decades. It is a low-cost, efficient way to establish the sod rotation. Midwest farmers overseed legume cover crops when planting corn, soybean or vegetable crops or before harvesting to maintain soil fertility.

Table 10.3 Effect of overseeding legume cover crops on corn yield and weed stand, 1981 (after Palada et.al. 1983)

Time of overseeding	Legume species	Grain yield (t/ha)	Weed reduction[a] (%)
35 DAP[b]	Medium red clover	7.30	76
	Hairy vetch	7.13	72
	Control (no overseeding)	7.49	--
47 DAP[c]	Medium red clover	6.96	40
	Hairy vetch	7.35	27
	Control (no overseeding)	7.13	--

a) The legume overseeding resulted in an average of 95% ground cover for both species.
b) DAP = days after planting corn, one cultivation prior to overseeding.
c) DAP = days after planting corn, two cultivations prior to overseeding.

In 1980, researchers at the Rodale Research Center examined the effects of legume species, time of overseeding and plant population on corn and soybean yields (Palada et al. 1983). Legumes overseeded during the first cultivation of the crop cycle had better germination and higher seedling emergence than those overseeded during the second cultivation. Overseeding during the first cultivation also provided significantly better ground cover than the second overseeding. These results suggest that during a dry summer, early overseeding provides excellent fall and winter ground cover. Legume cover crops did not reduce grain yield of corn and soybean (Table 10.3). Weed competition in both crops was also significantly reduced by overseeding.

Light level has a major influence on the survival and persistence of legume cover crops under row crop canopies. As the soybean canopy begins to close, the light intensity under the canopy decreases, suppressing sod growth. At full canopy, the sod is eliminated because light penetration under soybeans is low. Researchers at Rodale Research Center are trying to identify species that will fix nitrogen and control erosion through fall, winter and early spring. This legume cover crop could either be plowed down in the early spring before planting another summer crop, or it could be continued as a sod rotation into the next year. Legume species that appear to have real promise are red and white clover, Austrian winter pea and hairy vetch.

Legume Sod-Based Rotations

Legumes in rotation or as green manure are useful in controlling soil erosion and maintaining soil organic matter. A typical three- to six-year crop rotation common among organic farmers in the Midwest and Northeast states involves alfalfa or clover, corn, soybeans and small grains, with the number of years of alfalfa or clover increasing with increasing slope.

Well-inoculated legumes provide substantial nitrogen for the next planting of grain crops. For example, first year alfalfa yielding seven to 11 tons per hectare will furnish most of the nitrogen needs of a following corn crop with corn yield equal to or higher than that of continuous corn fertilized at 150 to 200 kg of nitrogen per hectare. A nitrogen fertility trial in corn conducted in 1979/80 at the Rodale Research Center showed no yield response to added fertilized nitrogen in fields that were organically managed and rotated with legume cover crops for more than five years. Legume/grass mixtures and clovers in which legumes are dominant are as effective in fixing nitrogen as a pure alfalfa stand producing the same amount of hay. However, alfalfa frequently yields more hay (Palada et al. 1983).

Sod Strip Intercropping

In strip intercropping, crops are grown simultaneously in different strips wide enough to permit independent cultivation, but narrow enough for two or more different crops to interact agronomically. The components can be a combination of row crops or a mixture of row crops and legume or grass sod. Using a legume sod is more advantageous from the standpoint of soil nitrogen. Sod strip intercropping may be limited to row crop production in sloping or hillside farms. These systems break the flow of water down the slope, reducing erosion substantially.

In 1978, researchers at the Rodale Research Center studied strip intercropping systems of red clover and short white clover with corn and soybean. Corn was planted in one-meter tilled strips at 40,000 plants per hectare using double rows. Soybeans were seeded in one-meter tilled strips at 250,000 plants per hectare. The check plots consisted of single rows planted in completely tilled soil, with no sod between the rows. Results showed that strip intercropping reduced corn yield by 17 percent to 34 percent, but did not affect soybean yield. Corn intercropped with short white clover had a slightly higher yield than with medium red clover. The researchers concluded that the choice of sod species may depend on how the legume is used other than as a cover crop. Red clover usually provides more biomass than short white clover, so it may be suitable for farmers who would make use of it for hay, silage or green mulch (Palada et al. 1983).

In another study on the effect of tillage width on corn, the researchers found that the highest yield (7.2 tons per hectare) was obtained from monoculture plots with single rows. Using a 0.75- or 1.5-meter tilled strip produced yields that were higher than other tillage widths. Yield reductions from these plots were only 8 percent and 16 percent, compared with 20 percent to more than 50 percent for other treatments. Rodale researchers concluded that monoculture corn produced more total dry matter than any of the combinations because of its superiority in yield. Although the intercrop system produced less total dry matter, the overall advantage is harvesting two feed crops in addition to reduced soil erosion and increased soil organic matter and nitrogen.

By manipulating tillage width, the system's total productivity can be adjusted to meet the grain and hay requirements of the farm. Tillage width can be adjusted as needed to fit available machinery with no adverse effects on soils and crop yields.

Vegetable/Living Mulch System

A living mulch system may be an economical way for vegetable growers to reduce soil erosion, increase organic matter and keep yields consistent with conventional systems. In 1978, Rodale researchers grew vegetables in existing grass and clover sod. The treatments consisted of red clover sod, bluegrass sod and completely tilled strips. Planting strips one-half-meter wide at two-meter spacing were prepared using a rotovator. Half of each treatment received 15 cm of alfalfa green chop while the other half was covered with black polyethylene mulch. The mulch was left undisturbed for one week so it could set back the existing untilled sod. Sweet corn and tomatoes were planted into the mulched strips. In completely tilled plots, half the rows between planting strips were seeded to short white clover and the other half were kept cultivated throughout the growing season. This was done to determine whether cultivating for weed control had any effect on the crop.

Production data showed that tomatoes grown in sod under a combination of mulch and tillage methods produced more fruit than those grown in a typical clean-cultivated field. Yield of tomatoes was 17 percent higher under alfalfa than under plastic mulch in the grass and clover sod strips, but not in the clean-cultivated field. Plants grown under black plastic mulch wilted and aborted flowers. These factors may have contributed to lower yields under plastic mulch.

The effect of mulch on sweet corn was the opposite of the effect on tomatoes. The most drastic differences occurred between mulch treatments, not between field treatments. More ears were harvested under alfalfa than under plastic mulch. Corn grown under plastic mulch shed pollen about two to four days before silking so that very little pollination occurred and yields declined.

Tilling the planting strips for both corn and tomato did not appear to affect their yields. However, tillage between rows of the clean-cultivated field decreased yields compared with strips seeded to white clover. This study suggests that both sweet corn and tomatoes can be adapted to living mulch systems, provided the living sod is set back by a mulch in the planting strip and enough soil moisture is continuously available. Competition with living sod was not a problem as long as the sod was properly managed.

11

Crop Rotation and Minimum Tillage

Crop rotation is a system in which different crops are grown in recurrent succession and in definite sequence on the same land (Page 1972). Several experiments lasting more than 100 years at the Agricultural Experiment Station at Rothamsted, England, and the Morrow plots at the Illinois Agricultural Experiment Station have provided considerable data on the effects of crop rotations. Evidence indicates that crop rotations influence plant production by affecting soil fertility and survival of plant pathogens, physical properties of soils, soil erosion, soil microbiology, nematodes, insects, mites, weeds, earthworms and phytotoxins (Sumner 1982).

BENEFITS OF ROTATING CROPS

Up until the 1950s, wheat and cotton yields in California depended on internal sources and recycling of nitrogen and organic matter. Nitrogen was obtained by rotating these crops with legumes. In fact, many farmers followed a fixed rotation system: a legume (alfalfa), a high-value crop (cotton) and a low-value grain (wheat). Alfalfa can produce up to 10 tons per hectare of dry matter and about 200 kg of nitrogen per hectare, sufficient to meet most of the nitrogen requirements of field and grain crops. In many parts of the United States corn belt, alfalfa can provide up to 50 percent savings on nitrogen costs for the first corn crop after alfalfa. Of course, during its year in rotation the alfalfa also produces high-quality feed for livestock.

Today, it is common in the corn belt to alternate the two major cash crops, corn and soybeans. Longer rotations of more than two years might include a year each with a small-grain crop and a legume/grass mixture for a hay crop. Economics is often the major determinant of crop selection.

Rotations can also suppress insects, weeds and diseases by effectively breaking the life cycles of pests. "Break" crops provide effective control of pests and diseases, the effectiveness increasing with the length or frequency of breaks. In most cases, a one-year break is sufficient to provide control, but this depends on environmental conditions and on the particular pathogen or insect species (Bullen 1967; Table 11.1).

Table 11.1 Effect of crop rotation of corn on insect populations or potential damage (after Metcalf and Luckmann, 1975)

	Corn rotation		
	None	Soybeans	Pasture and hay crops
Seed corn beetles	0	0	+[a]
Seed corn maggot	0	0	+
Wireworms	−	−	+
White grubs	−	+	+
Corn rot aphid	−	−	+
Grape colaspis	−	−	+
Northern corn rootworm	+	−	−
Western corn rootworm	+	−	−
Southern corn rootworm	0	0	0
Black cutworm	0	+	0
Billbug	−	−	+
Slugs	−	−	0
Thrips	0	?	+
Mites	0	0	0
European corn borer	0	0	0
Southwestern corn borer	0	0	0
Corn earworm	0	0	0
Fall armyworm	0	0	0
True armyworm	0	0	+
Chinch bug	0	0	+
Corn leaf aphid	0	0	0
Totals			
+	2	2	10
−	6	7	2
0	13	11	9
?	0	1	0

[a] + means the practice will increase the population or damage from that insect; − means it will reduce the population or damage; 0 means no effect; ? means effect unknown.

The sequence of crops within a rotation may be critical since some crops yield better or worse depending on the crop they follow. Most experiments have documented the detrimental effects of continuous cropping of corn and small grains on organic matter and nitrogen in unfertilized plots. Sorghum is a notoriously hard crop to follow. Yields of almost any crop after sorghum will be lower than after corn, soybeans or wheat. It has been suggested that sorghum's effect on succeeding crops is due to the high carbohydrate content of sorghum's roots. The decomposition of the roots stimulates soil microbial growth and "ties up" nitrogen and other nutrients in the soil microflora, a phenomenon known as immobilization. In other cases, the effect of one crop on the next may relate to chemicals left in the soil or generated by decomposition of the crop residues. Wheat residues, for example, have been shown to inhibit the growth of several different crops that might follow. The allelochemicals are thought to be produced during decomposition of the residues by certain soil microbes.

In turn, research has shown that plots that included a legume as a green manure increased yields of those crops. The benefits of green manuring were obtained by storing organic matter and nutrients in the soil improvement crops and releasing the nutrients by decomposition of the organic matter when they were of the most benefit to the following crop. The most important contribution of winter legume cover crops, especially on sandy soils, was increased nitrogen (Doll and Link 1957). Throughout the U.S., especially in areas with reasonably long frost-free seasons, a number of rotations have been developed (such as wheat/soybean/winter legume/corn, wheat/corn for silage; annual ryegrass grazed and allowed to reseed/soybeans; winter small grain/interseeded into summer crops).

With the availability of inorganic fertilizers in modern agriculture, the need for crop rotation purely from the standpoint of soil fertility has diminished. The greatly increased supply of chemical nitrogen in the U.S. in the 1950s prompted much interest in continuous cropping. As prices of energy and nitrogen fertilizer increase, rotations may once again become cost effective, and substantial energy savings will surely ensue. Heichel (1978) showed that corn-based crop rotations incorporating grain and forage legumes reduce the demand for energy. Compared with continuous cropping, the fossil energy flux in rotations is reduced as much as 45 percent (Table 11.2).

Obviously, the particular cropping sequence used in a rotation will vary with the climate, tradition, economics and other factors. It should be expected, however, that crop rotations will broaden the economic base of the agricultural enterprise, spread labor demands more evenly through the year and allow production of

Table 11.2 Intensity and efficiency of energy use in continuous corn compared with crop rotations incorporating grain and forage legumes (Heichel 1978)

	Rotation					
	1	2	3	4	5	6
	Continuous corn	2 Corn-soybeans	2 Corn-oats-2 alfalfa	3 Corn-3 soybean-wheat-3 alfalfa	2 Corn-alfalfa	Corn-soybeans-vetch
Fossil energy flux (Mcal/a-day)	17.4	12.9	10.7	9.7	11.1	8.9
Crop yield (lbs dry matter/a)	7,767	6,216	7,337	6,150	6,645	5,200
Crop energy yield/fossil energy flux	6.1	6.7	7.8	8.3	8.1	8.2

high-value crops, thereby increasing income opportunities (Briggs and Courtney 1985).

MINIMUM TILLAGE SYSTEMS

Minimum tillage is any tillage system that reduces soil loss and conserves soil moisture, as compared with unridged or clean tillage (Mueller et al. 1981). Under this system unincorporated plant residues are left on the soil, and its surface is left as rough as possible.

Crop production using no-tillage methods has been shown to reduce material and energy inputs and, perhaps more importantly, abate soil erosion. No-tillage systems also improve the scheduling and reliability of farm operations since many weather-related constraints are alleviated. Crops grown by these practices can usually be planted, treated for weeds and harvested when tilled fields would be too muddy to enter.

Other advantages include moisture conservation, reduced soil compaction and an increase in multiple-cropping potential. Furthermore, crop yields from no-tillage systems frequently equal or exceed the yields from conventional methods (Phillips and Phillips 1984). A USDA study estimated that by the year 2000, as much as 65 percent of the U.S. acreage of field grains—wheat,

rye and soybeans—will be produced by reduced tillage methods (Phillips et al. 1980).
The no-till system causes very little soil disturbance. The one-pass tillage and planting operation tills a slot approximately five cm wide for seed placement. The slot is usually opened with a fluted colter placed ahead of the planter unit. Since there is no soil disturbance, more than 95 percent of the residue is left on the surface.

Effects on Soil Characteristics and Plant Growth

Soil moisture. Tillage systems leaving 50 percent or more of the soil surface residue covered after planting generally increase soil moisture throughout the season due to increased filtration and decreased evaporation. In areas with low annual rainfall and on soils with low water-holding capacity, the added water should increase yield potential. On poorly drained soils in northern latitudes the extra water may delay planting and reduce yield potential (Sprague and Triplett 1986).

Soil temperature. Several studies have shown that increased surface residue slows the rate of soil warming in the spring, therefore delaying germination, emergence and early growth of crops, especially in the northern United States. However, this could be a benefit in the southern United States and in more tropical climates. Different types of tillage systems leave varying amounts of residue on the surface and, as a result, soil temperatures will vary among them. Differences in soil temperature between no-till and conventional practices can vary from 1° to 4° C.

Soil fertility. Because of the increased residue and reduced tillage, minimum tillage systems produce different levels of moisture, temperature, organic matter content and rate of decomposition and microbial population. All these factors influence the availability of nutrients and thus the need for fertilizer. Leaving residues on the surface causes organic matter to build up near the soil surface, with positive effects on soil physical properties. Unfortunately, from the studies conducted thus far, researchers have not reached any conclusions as to whether nitrogen fertilizer programs must be changed for minimim tillage systems.

Some evidence suggests that surface residues left the first year after the adoption of no-tillage will exert a strong demand on available nitrogen and may cause deficiencies or at least lower nitrogen availability. However, after several years of minimum tillage, the system stabilizes and nitrogen fertility no longer varies from conventional tillage. Phosphorus seems to have equal or greater availability under no-till compared with the conventional system, regardless of whether the fertilizer was broadcast or banded. This phenomenon occurs despite the fact that broadcast phosphorous accumulates in the top centimeter of the soil under no-till because of the lack of incorporation and movement through the soil profile. Possibly the residues on the surface allow sufficient moisture for root growth and uptake of phosphorous nutrients.

There is disagreement about the availability of potassium under no-tillage. Some researchers have found decreased available potassium, especially under some wet and cold conditions, while others have reported no deficiency. Further research should clarify these conflicting views.

Soil acidity. Soil acidity becomes a greater factor under no-tillage. Of the secondary and microelements, magnesium is little affected and sulfur is likely to be less available from the soil organic matter. Zinc tends to be more available due to higher organic matter content and lower pH. In general, soil fertility under no-tillage is strongly influenced by the interacting effects of increased soil mosture, high levels of slowly decomposing organic matter in the soil, higher acidity and lower temperatures in the spring (Sprague and Triplett 1986).

Effects on Pests

Weed control. Conservation tillage systems depend on heavy applications of herbicides. Usually the maximum recommended herbicide rate is used in corn because of the surface accumulation of weed seeds, which potentially exert greater weed pressure than under conventional tillage. In addition, surface residue intercepts and inactivates part of the applied herbicide.

Eliminating tillage causes shifts in weed species. Perennials readily controlled by tillage become established and persist in untilled fields. Weeds botanically related to the crop and others that escape control often increase, becoming major problems. A classic example is the increase of fall panicum in corn after repeated applications of atrazine to control annual weeds in no-till systems (Sprague and Triplett 1986).

Disease control. Alterations of the microclimate due to surface residues may retard, enhance or have no effect on plant diseases. The degree of influence on plant diseases by residue generally relates to the amount of residue remaining after planting.

Surface residues may affect plant diseases in several ways. They provide a habitat for overwintering (survival), growth and multiplication of plant pathogens, particularly fungal and bacterial pathogens. There are many plant pathogens that overwinter best in surface residues because they are protected from the environment and other microorganisms. Surface tillage increases the chances of epidemics caused by such pathogens.

During a seven-year study, foliar diseases were never observed to be a problem on grain sorghum or wheat grown under minimim tillage in Nebraska (Doupnik and Boosalis 1980). The incidence of stalk rot of grain sorghum, a stress disease caused by Fusarium moniliforme, was dramatically reduced under no-tillage compared with that under conventional tillage. Its incidence was reduced from 39 percent to 11 percent, and yield was increased (Sprague and Triplett 1986). The increased soil moisture storage and the lower, more constant soil temperature associated with minimum tillage are undoubtedly two major factors accounting for the lower incidence of stalk rot in corn. Under these more favorable growing conditions, the plants are less vulnerable to this fungus. On the contrary, in Wisconsin, eyespot (a leaf disease of corn) has been shown to be more severe on corn grown under no-tillage.

Crop rotation is especially important for controlling diseases with surface tillage. Planting a crop in its own residue from the previous season (monoculture) without a fallow period is more likely to increase certain plant diseases than a system in which a crop is planted into the residue of an unrelated crop. Another way to reduce diseases associated with reduced tillage is to rotate tillage systems. Inclusion of tillage rotation with crop rotation is an excellent method of disease management. This could be done in a manner to allow retention of 20 to 30 percent of the surface residue, thus providing the benefits of surface tillage while reducing the potential of disease outbreak.

Soil-borne fungal diseases associated with surface tillage may be decreased by the kind, amount and time of fertilizer application. Applying ammonium sulfate in the spring controlled take-all of wheat, whereas fall application of nitrogen for spring-seeded wheat increased take-all.

Insect dynamics. Entomologists working in no-till agriculture have found that the mulch-litter layer of no-tillage soil provides a favorable microhabitat for some insects that attack corn, such as army worms, black cutworm and stalk borers (House and Stinner 1983). The loss of reliable mechanical destruction methods in no-tillage corn increases the survival of insect pests inhabiting crop residue or residing on or near the soil surface. The greatest hazard for pest infestation occurs in the seed and seedling stages from subterranean insect pests. Two pest trends are often associated with no-tillage systems: (1) the level of pest activity is related to the previous crop type, and (2) no-tillage systems commonly support a higher diversity of pest insects than conventional tillage systems. Most approaches to solving pest problems in no-tillage systems have been largely symptomatic. Almost exclusive reliance has been placed on broad spectrum insecticides, and little research has been devoted to developing cultural and biological methods of pest suppression and prevention.

Recently, investigators in Georgia have reported entomologically beneficial aspects inherent to no-tillage systems (House and Stinner 1983). For example, the lesser corn stalk borer Elasmopalpus lignosellus feeds preferentially on grain stalk residues in no-tillage corn systems. Thus, infestations of the lesser corn-stalk borer are deterred. In tropical Costa Rica, Shenk and Saunders (1983) found that incidence of the fall armyworm (Spodoptera frugiperda) and the leaf beetle (Diabrotica balteata) was much greater in plowed maize plots than in no-till plots. In north Georgia soybeans, the abundance and diversity of carabid beetles are often several times higher in no-tillage than in conventional tillage soybeans. The surface litter and weeds on no-tillage usually provide the predatory carabid and spider fauna with greater food resources as well as protection from unfavorable climatic conditions (House and Stinner 1983). Their control of some seed and seedling pests can be substantial. Increased moisture and reduced temperature can enhance development of insect pathogens, as it was observed with Rhabditoid nematodes in no-till sorghum (Sprague and Triplett 1986).

Crop Yields

A number of yield-comparing studies have revealed that, in northern climates on fine-textured soils, no-till systems will probably yield lower than conventional systems. This is attributed to reduced soil temperatures (1° to 4° C) and a shorter growing season. Minimum tillage systems may produce greater yields on droughty soils and in well-drained soils, or in more southerly climates. No-till systems have a long-term yield advantage on soils highly susceptible to erosion.

Energy Requirements

Less energy is required for tillage operations in many no-till systems. Since plowing, disking and other trips over the field are eliminated, these systems result in 34 percent to 76 percent reductions in fuel for tillage operations. However, the requirement for additional herbicides in no-tillage systems may offset some of these gains. In general, though, total production costs for corn in the midwest U.S. rise slightly with the intensity of tillage.

Research on crop rotations including crops that leave residues with allelopathic activity against certain weeds (see Chapter 14) is clearly warranted to reduce herbicide use in no-tillage systems. Many annual broadleaf weeds are suppressed if mulches, especially small grain cover crops, are left on the soil surface (Putnam and DeFrank 1983). Fall-dessicated "Balboa" rye and "Tecumseh" wheat used in rotations can greatly reduce weed biomass in the next growing season by inhibiting germination and growth of several weed species.

12

Agroforestry Systems

John G. Farrell

Agroforestry is the generic name used to describe an old and widely practiced land use system in which trees are combined spatially and/or temporally with agricultural crops and/or animals. However, only recently have modern concepts of agroforestry been developed, and to date no universally acceptable definition has evolved, although many have been suggested (Agroforestry Systems, Vol. 1, pp. 7-12 and Wiersum 1981).

CHARACTERISTICS OF AGROFORESTRY

Agroforestry incorporates four characteristics:

1. Structure: Unlike modern agriculture and forestry, agroforestry combines trees, crops and animals. In the past, agriculturalists rarely considered trees useful on farmland, while foresters have regarded forests simply as preserves for growing trees (Nair 1983). Yet for centuries traditional farmers have provided for their basic needs by raising food crops, trees and animals together.

2. Sustainability: Agroforestry aims to optimize the beneficial effects of interactions between woody species and crops or animals. By using natural ecosystems as models and applying their ecological features to the agricultural system, it is hoped that long-term productivity can be maintained without degrading the land. This is particularly important considering the current application of agroforestry in areas of marginal land quality and low availability of inputs.

3. Increased productivity: By enhancing complementary relations among farm components, and with improved growing conditions and efficient use of natural resources (space, soil, water, light), production is expected to be greater in agroforestry

systems than in conventional land use systems.

4.   Socioeconomic/cultural adaptability: Although agroforestry is appropriate to a wide range of farm sizes and socioeconomic conditions, its potential has been particularly recognized for small farmers in poor, marginal areas of the tropics and subtropics. Considering that peasants are usually unable to adopt modern, high-cost agricultural technologies, have been bypassed by agricultural research and have no discernible political or social power, agroforestry is particularly adapted to the circumstances of small farmers.

CLASSIFICATION OF AGROFORESTRY SYSTEMS

Several criteria can be used to classify agroforestry systems and practices (Nair 1985). Most commonly used are the system's structure (composition and arrangement of components), function, socioeconomic scale and level of managment, and ecological spread. Structurally, the system can be grouped as agrisilviculture (crops, including tree/shrub crops, and trees), silvopastoral (pasture/animals plus trees) and agrosilvopastoral (crops plus pasture/animals plus trees). Other agroforestry systems can be specified, such as apiculture with trees, aquaculture in mangrove areas, multipurpose tree lots and so on. The components can be arranged in time or space, and several terms are used to denote the various arrangements. Functional basis refers to the main output and role of components, especially the woody ones. These can be productive functions (production of basic needs, such as food, fodder, fuelwood, other products) and protective roles (soil conservation, soil fertility improvement, protection offered by windbreaks and shelterbelts).

On an ecological basis, systems can be grouped for any defined agroecological zone such as lowland humid tropics, arid and semi-arid tropics, tropical highlands and so on. The socioeconomic scale of production and level of management of the systems can be used as the criteria to designate systems as commercial, intermediate or subsistence. Each of these criteria have merits and applicability in specific situations, but they also have limitations, so no single classification scheme can be considered universally applicable. Classification will depend on the purpose for which it is intended.

The Potential Role of Trees

Trees are generally underused in agriculture, and although much has been written of their virtues (Smith 1953, Douglas and Hart 1976, MacDaniels and Lieberman 1979), their potential has gone relatively unexplored. By virtue of their form and growth habit, trees influence other components of the agricultural system (Figure 12.1). Their large canopies affect solar radiation, precipitation and air movement, while their extensive root systems fill large volumes of soil. Absorption of water and nutrients and the redistribution of nutrients as leaf litter, as well as the disruptive movement of the roots and possible root fungal/bacterial associations, can also alter the growing environment.

Trees can enhance the productivity of a given agroecosystem by influencing soil characteristics, microclimate, hydrology and other associated biological components.

Soil characteristics. Trees may affect the nutrient status of the soil by exploiting the deeper mineral reserves in the parent rock, and by retrieving leached nutrients and depositing them on the surface as leaf litter. This organic matter increases the soil humus content, which in turn increases its cation exchange capacity and decreases nutrient losses. The added organic matter also moderates extreme soil reactions (pH) and the consequent availability of both essential nutrients and toxic elements. Since nitrogen, phosphorus and sulfur are primarily held in organic form, plenty of organic matter is especially important to make them available. The association of trees with nitrogen-fixing bacteria and mycorrhizae will also increase available nutrient levels. Microorganism activity tends to increase under trees because of increased organic matter (improved food supply) and growing environment (soil temperature and moisture).

A study conducted to evaluate the role of trees in traditional farming systems of Central Mexico (Farrell 1984) illustrates the potential influence of trees on soil fertility. Surface soil properties were measured at increasing distances from two tree species, capulin (Prunus capuli) and sabina (Juniperus deppeana), found within selected crop fields. Higher values of all properties measured were found under the capulin canopies, and a decreasing gradient was observed with increasing distance from the trees. Available phosphorous increased fourfold to sevenfold under the trees (Figure 12.2) and total carbon and potassium increased two- to threefold; nitrogen, calcium and magnesium increased one-and-a-half- to threefold and cation exchange capacity increased one-and-a-half- to twofold. Soil pH was also found to be higher under the canopies. This spatial pattern was attributed primarily to the redistribution of nutrients with litter fall and the accumulation of organic matter near the capulin trees.

Trees may also enhance the physical properties of soil, the most important being soil structure. Structure improves as a result of increased organic matter (leaves and roots), and the disruptive action of the tree's roots and microorganism activity,

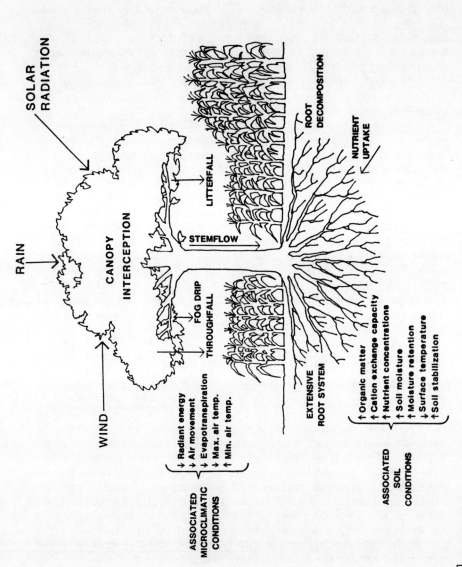

Figure 12.1
The influence of trees in Tlaxcala, Mexico, on the growing environment of maize (Farrell 1984).

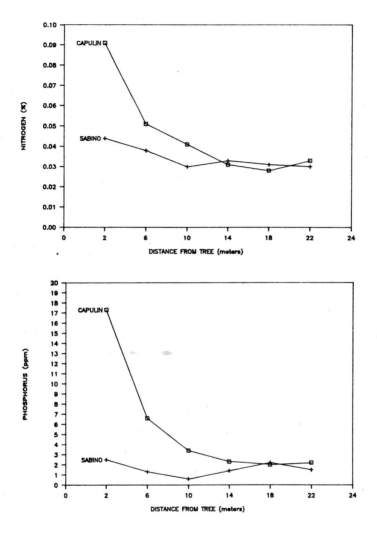

Fig. 12.2 Change in surface soil nitrogen and phosphorus with increasing distance from individual capulin and sabino trees (Farrell 1984)

all of which aid in developing more stable soil aggregates. Soil temperature is moderated by the shade and litter cover.

The role trees may play in soil protection is well recognized. In addition to reducing wind velocities, the tree canopy diffuses the impact of rain drops hitting the soil surface. The litter layer covering the soil and its improved structure also help reduce surface erosion. The penetrating root system of trees serves an important function in stabilizing the soil, especially on steep slopes.

Microclimate. Trees moderate temperature changes, resulting in lower maximum and higher minimum temperatures under trees as compared with open areas. Lowered temperature and reduced air movement due to tree canopy reduces the evaporation rate. Greater relative humidity may also be found under trees compared with open sites.

Hydrology. The water balance of a given microsite, farm or region is influenced by both structural and functional characteristics of trees. To varying degrees, depending on canopy density and leaf characteristics, precipitation either passes through to the soil surface, is intercepted and evaporates or is redistributed to the base of the trunk with stemflow. Air moisture may also be collected by tree canopies and deposited as internal precipitation (fog drip), a potentially significant source of water in moist foggy areas. As a result of improved soil structure and the presence of a litter layer, the water that does reach the ground is used more efficiently due to increased infiltration and permeability, and evaporation and surface runoff are reduced. On a larger scale, particularly in areas prone to flooding, trees may reduce subterranean water discharge.

Associated Biological Components. Crop and non-crop plants, insects and soil organisms can benefit from the presence of compatible trees as a result of the various characteristics discussed above. Although the specific mechanisms are little understood, they generally involve a more benign microclimate; favorable soil temperature, moisture regime and organic matter status; and increased availability of nutrients, as well as their efficient use and recycling.

Productive role. Trees produce a number of products important to both humans and animals. In addition to food and forage, they provide wood products, by-products such as oils and tannins and medicinal products. For example, the black locust (Robinia pseudoacacia) is an important honey plant, fixes nitrogen and is a source of durable fenceposts. Leucaena, another nitrogen-fixing legume, is valuable as poultry and cattle food in the tropics because of its high vitamin and protein content. It is also a primary souce of firewood (NAS 1977). Tree crops can also supplement grain production. Species such as chestnut (Castanea), carob (Ceratonia) and honey locust (Gleditsia) have a higher food value in proteins, carbohydrates and fats than some conventional grains (Smith 1953), and also grow on marginal land without cultivation.

DESIGNING AGROFORESTRY SYSTEMS

Natural ecosystems can be useful as models for designing sustainable agricultural systems. The most conspicuous feature of natural forests is their multistoried organization of trees, shrubs, forbs and fungi, each using different levels of energy and resources and each contributing to the functioning of the entire system. These layers lessen the mechanical impact of raindrops hitting the soil surface and reduce the amount of direct sunlight reaching the ground, thereby minimizing the potential for soil loss, reducing evaporation and slowing the rate of organic matter decomposition. There is generally little wind at ground level. On the soil surface, decaying plant litter provides a protective cover and a source of nutrients to be recycled.

All these conditions create an ideal environment for microflora and fauna, insects and earthworms that promote the decay and incorporation of organic matter into the soil, creating good soil structure, which in turn enhances aeration and water infiltration. Insects that are potentially harmful to the vegetation are kept in check by resident predators and parasites. There is also multilayering below the surface, where the roots of various plant forms use different soil volumes. Thus, nutrients that are leached below the rooting zone of the smaller vegetation are intercepted by the deeply penetrating roots of trees and returned to the surface as leaf litter.

The main goal in designing an agroforestry system is to enhance the fundamental ecological features of the forest, so an understanding of these processes in a natural system is essential. Most of the principles outlined in Chapter 7 can be applied to designing agroforestry systems, particularly Hart's ideas (1978) about designing crop sequences analogous to natural succession. This approach can be very useful in regions lacking natural vegetation, where successional models from ecologically homologous areas can be initiated. Oldeman (1981) proposed the "transformation" concept as another design option. Complementary to the analog approach, its base is the structural analysis of collective units (eco-units). Transformation may be brought about by replacing wild species with useful species fulfilling the same functional and structural niche as their wild precedents. This process transforms the structure of the natural system while maintaining its beneficial properties.

In situations where a fully forested area is not suitable for a farm, trees may be combined with crops and animals in other ways to enhance the desired functional relationships. Wiersum (1981) and Combe and Budowski (1979) have outlined these practices in their attempts to develop a classification system for agroforestry techniques.

Plant Arrangements

A number of factors should be considered in arranging component plant species in space and time. These may include the cultural requirements of component species when grown together, their growth form (both above and below ground) and phenology, management requirements for the entire system and the need for additional actions such as soil conservation or microclimate amelioration. Thus, plant arrangement patterns are site specific. Possible patterns include (Nair 1983):

1. Intercropping tree species with annual agricultural crops, planting both herbaceous and woody species simultaneously (or in the same season). The spacing of woody species will vary considerably, but generally they will be spaced more widely in drier regions. This scheme can also be applied to agricultural plantation crops such as rubber and oil palm.

2. Clearing strips about one meter wide in primary or secondary forests at convenient intervals and planting shade-tolerant perennial agricultural species such as cacao. Subsequently, as the planted species grow up, the forest vegetation will be selectively thinned, and in about five years there will be a two- or three-layer canopy consisting of the perennial agricultural species and the selected forestry species.

3. Introducing management practices such as thinning and pruning to allow more light to penetrate to the plantation floor, and planting selected agricultural species between the rows of trees. The extent of thinning or pruning will depend upon the tree density, canopy structure and so forth.

4. In hilly areas, selected tree species can be planted in lines across the slope (along the contour) in different planting arrangements (single rows, double rows, alternate rows), with varying distances between rows; soil-binding grasses can be established between the trees along the contours. The area between the rows can be used for agricultural species.

5. Close-planting multipurpose trees around plots of agricultural fields. The trees will form live fences and windbreaks, provide fodder and fuel and mark the boundaries of agricultural plots. The scheme is particularly suitable for extensive land use areas.

6. Interspersing intensively managed agricultural areas with trees, in a regular or haphazard manner. The system is popular in smallholder farming in Asia, the Pacific, Africa and South America.

Examples of Agroforestry

Home gardens in the tropics are one of the classic examples of agroforestry. Home gardens are a highly efficient form of land use, incorporating a variety of crops with different growth habits. The result is a structure similar to tropical forests, with diverse species and a layered configuration. Some of the most refined gardens can be found in West Java, as described in Chapter 6.

Intensive intercropping with plantation crops such as coconut, cacao, coffee and rubber is another agroforestry technique. In India, crops such as black pepper, cacao and pineapple are grown under the coconut, using the available light as well as a greater percentage of soil volume (Nair 1979). Coffee, tea and cacao are traditionally grown under one or two strata of shade trees; these often are nitrogen-fixing legumes that also produce valuable wood products.

In semi-arid and arid parts of the world, the use of multipurpose trees mixed with crops or as part of pastoral systems is the dominant agroforestry practice. Species such as Acacia and Prosopis are valuable, not only for their wood products and forage, but also for their soil-enriching qualities. The unique phenology of Acacia albida (leafless during the rainy season) makes it an ideal component of the sorghum- and millet-producing regions of West Africa and the Sahelian zone.

Similar uses of trees have been described in Mexico (Wilken 1977), where farmers encourage the growth of native leguminous trees in cultivated fields. From Puebla and Tehuacan south through Oaxaca, farms with light to moderately dense stands of mesquite (Prosopis spp.), guaje (Leucaena esculenta), and guamuchil (Pithecellobium spp.) are a familiar site. Stand density varies from fields with only a few trees to virtual forests with crops planted beneath.

A slightly different practice is found near Ostuncalco, Guatemela, where rigorously pruned sauco (Sambucus mexicana) stumps dot maize and potato fields. Leaves and small branches are removed annually, scattered around individual crop plants, then chopped and interred with broad hoes. Local farmers claim that crop quality and yields in the sandy volcanic soils of this region depend upon these annual applications of sauco leaves.

Trees are integrated with farm animals in many areas. They range from small animals confined in home gardens in the tropics, to livestock grazing in orchards in Chile (Altieri and Farrell 1984) to livestock grazing in forest plantations in New Zealand (Tustin et al. 1979) or the southeastern United States (Lewis et al. 1984).

NOTES

1. Agroecology Program, University of California, Santa Cruz.

PART FOUR

Ecological Management of Insect Pests, Pathogens and Weeds

13

Pest Management

Some pest management programs have been unjustifiably slow in putting ecologically based theory into practice. The lack of training in "holistic thinking," the sense of short-range urgency in applied "practical" research, and the "need" to defend a particular discipline from intrusion of "others" can all inhibit the integration of theory and application.

Unfortunately, pest control research has been too subservient to pressure groups in the U.S. Congress, large farmer associations and particularly, the pesticide industry. Since biological methods of pest suppression do not lend themselves readily to agribusiness systems for large-scale manufacturing and marketing, as do conventional pesticides, it is evident that private enterprises are reluctant to incur the expense of their development.

Furthermore, in some cases bio-environmental strategies in crop protection involve yield stabilization rather than maximization. In ecologically managed systems, productivity on a crop-unit basis might be reduced, but other desirable environmental features, such as the multiple-use capacity of the habitat, are enhanced. The net result appears to be greater resource diversity and overall biological stability, though it could be argued that these advantages are not substantial and tangible enough to be justified from the commodity yield point of view. Nevertheless, energy shortages and economic inflation will probably show that short-term financial gain should no longer be the primary driving force in agriculture. Possibly, energy conservation and environmental quality will assume such a role.

AGROECOLOGY AND PEST MANAGEMENT

There are many opportunities for multidisciplinary research in pest management. For example, the biogeographic region, rather than the single homogeneous field, may be the appropriate unit for pest management research. The agroecosystem should be conceived of as an area large enough to include those uncultivated areas that influence crops through intercommunity exchanges of organisms, materials and energy (Rabb 1978). The influence of hedgerows, forest edges and crop field borders on the fauna of farm systems must be ascertained. The interchange of organisms among regions and even among crops must be investigated, as must the role of uncultivated habitats in the overwintering ecology of pest and beneficial organisms.

Natural ecosystems can be regarded as models for pest management strategies in agroecosystems. Some rural societies simulate forest conditions in their farms to obtain the beneficial effects of forest structures. Farmers in Central America imitate the structure and species diversity of tropical forests by planting a variety of crops with different growth habits. By keeping diversity at the highest possible level, small farmers have minimized the threat of unstable conditions (such as pests) while obtaining a stable source of income and nutrition and maximizing returns under low levels of technology. Capitalizing on knowledge of beneficial plant associations through multiple cropping research can lead to the creation of systems that use resources, improve overall productivity, buffer against pest epidemics and at the same time conserve the ecosystem.

Understanding interactions among plants, insect herbivores and their natural enemies can provide several clues to improve biological control systems. Plants can have several direct and indirect, positive and negative effects not only on the herbivores but also on their natural enemies. There are numerous ways in which plants can directly affect the performance of entomophagous arthropods. Chemical and physical cues emitted by the plants are essential for effective location of the host habitat by entomophages (Monteith 1960).

For example, eggs of Heliothis spp. and Plusiinae pests suffer higher parasitization by Trichogramma wasps on tomato plants than on okra, corn, sorghum, tobacco and clover. This suggests that the host plant habitat probably affects the degree of parasitization obtained from augmentative releases of Trichogramma, regardless of the target insect pest. However, no attempts have been made to use plants and their volatiles to improve the habitat and host selection of parasitic insects. Recently, researchers have found that parasitization of corn earworm eggs by wild Trichogramma wasps can be enhanced by increasing the chemical diversity of crop fields through the application of extracts of the weed Amaranthus sp. (Altieri et al. 1983b).

Integrated Pest Management

Integrated pest management (IPM) should be oriented to prevent outbreaks by improving the stability of the crop systems rather than coping with pest problems after they occur. Currently, the issue in pest management is to design systems that suppress a complex of pests while achieving maximum yield and quality, and minimum environmental damage. These objectives may appear to conflict, and when yield and market quality are overemphasized, they usually do. However, the conflict can be avoided when IPM systems are coordinated with more broadly related systems of land and water, resource conservation, environmental protection and socioeconomic development. IPM systems should be designed to balance pests and beneficial organisms based on known economic, social and ecological consequences.

An equilibrium of the crop fauna can be established by organizing the vegetational diversity within and without the target crop fields. Providing the right kind of plant diversity throughout the year and manipulating time of planting, size of fields and species composition of crop field borders can make habitats and food resources continually available for populations of beneficial organisms and make habitats less favorable for pests (Table 13.1). (Litsinger and Moody 1976, Altieri and Whitcomb 1979).

Vegetational Diversity and Pest Problems

Most entomologists, plant pathologists and weed scientists agree that the intensification that has accompanied the growth of agriculture has promoted several practices that favor insect pests, weeds and diseases. These include (Zadoks and Schein 1979, Pimentel and Goodman 1978):

o Enlargement of fields, resulting in extensive monocultures or short rotational patterns of low species diversity

o Aggregation of fields of similar species and/or varieties, decreasing mosaic diversity at the regional level

o Increase in the density of host crop plants by adopting crop spacings that encourage pest outbreaks and epidemics

o Increase in the uniformity of host populations, thereby decreasing genetic diversity. When the genetic makeup of a crop is altered to increase its yield, with little attention to pest attack, natural resistance to insects and pathogens can be greatly reduced.

Plant pathologists recognize the following patterns of disease behavior in monocultures (Zadoks and Schein 1979): (1) disease

Table 13.1 Vegetational arrangement of crops and other plants in time and space evaluated as to the potential development of pest problems (after Litsinger and Moody 1976)

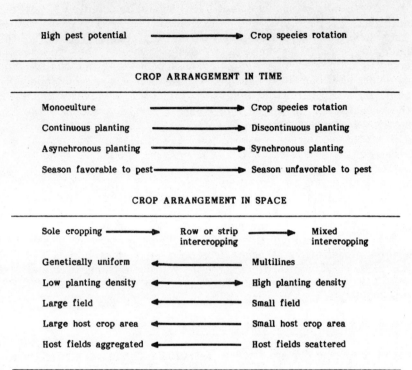

increases to maximum intensity and severity, remaining for the duration of the monoculture; (2) disease increases to a limited extent with moderate intensity, maintained at plateau; (3) disease shows minimal or undetectable development throughout the monoculture; (4) disease exhibits a variable intensity of development, in erratic cycles; and (5) disease increases to a peak of maximum intensity followed by a period of declining severity.

With insects these patterns have not been clearly defined. However, the analysis of Andow (1983) and Altieri and Letourneau (1982) indicate that insect pest abundance often increases with extended periods of monoculture, destruction of woodlands and hedgerows, disproportionate increases in crop acreage, replacement of diversified farming or low-maintenance crops and replacement of natural forests by annual crops. Andow (1983) mentions four examples (two in cotton and two in wheat) in which increasing monocultures eliminated alternative host plants, so large populations of multivoltine, polyphagous herbivores (e.g., Heliothis zea, Bemisia tabaci and Oscinella spp.) were unable to build up. Present evidence supports Andow's hypothesis, that monophagous, less vagile pests are good candidates for outbreaks depending upon increasing monoculture. Other agricultural practices such as fertilization, irrigation and pesticide applications can make plants more or less susceptible to pest and disease attack.

Researchers searching for an ecological approach to pest control envision the restoration of plant diversity in agriculture. They hope that by adding selective diversity to crop systems, it will be possible to capture for agroecosystems some of the stable properties of natural communities (Root 1973). Several researchers have explored the effects of increased diversity on pest management (Litsinger and Moody 1976, Perrin 1977, Cromartie 1981, Altieri and Letourneau 1982).

Although diversifying agroecosystems does not automatically reduce pest problems, about 53 percent of the insect pests studied in 150 experiments have shown a clear reduction in pest incidence in more diversified systems compared with corresponding monocultures (Risch et al. 1983). Only 18 percent of the pest insects studied increased in the diverse systems. Despite this striking experimental evidence, only a few attempts have been made in the developed world to increase diversity for the purpose of enhancing pest management.

Monocultures in the United States

Monocultures have increased dramatically in the U.S. both spatially, in that the land devoted to single crops has expanded within a geographic area, and temporally, through the year-to-year production of the same species on the same land. Although it is difficult to quantify the extent of monocultures using present information, a task force on spatial heterogeneity in agricultural landscapes reported that crop diversity per unit of arable land has decreased and croplands have become increasingly concentrated (USDA 1973). These trends are particularly evident in the corn belt, Mississippi delta, Red River valley, Texas high plains, California irrigation areas, southern Florida, and the Kansas/Oklahoma winter wheat belt.

The state of Illinois was the subject of a pilot analysis of trends in agricultural diversity over 32 years (USDA 1973). Studying corn, soybeans, wheat, oats, hay and plowland pasture, USDA scientists applied the Shannon-Weaver index to nine crop districts for year-to-year comparisons of diversity. In seven districts the diversity fell precipitously beginning in the mid-1950s. Specialization of corn and soybean production (not the diverse species assemblages that make up hay or pastureland) accounted for the increase in monoculture and aggregation of croplands in most of the studied districts.

Political and economic forces usually influence the trend to devote large areas to monoculture. Concentration of directed research on particular commodities increases production technology and profitability for certain commodities, thus contributing to monoculture. Capital-intensive, directed agricultural research products tend to be large-scale and to benefit large monoculture farmers. Also, because of economies of scale, the concentration of managerial expertise, mechanization and marketing all support the trend toward monoculture practices (Buttel 1980b).

THE IMPORTANCE OF DIVERSITY

Diversity and Insect Populations

Exposed fields and concentrations of a single crop species open the way for pest infestations by providing concentrated resources and uniform physical conditions that encourage insect invasions (Root 1973). The abundance and effectiveness of predators are reduced because these simplified environments provide inadequate alternative sources of food, shelter, breeding sites and other environmental factors (van den Bosch and Telford 1964). Herbivorous insect pests are more likely to colonize and remain longer on crop hosts that are concentrated because the entire life requirements of the pests are met in these simple environments (Root 1973). As a result, populations of specialized pests attain economically undesirable levels.

Two hypotheses explain pest reduction in polycultures (Root 1973; Altieri et al. 1978, Bach 1980, Risch 1983, Altieri and Liebman 1986 and 1988). The first, the natural enemies hypothesis, predicts greater mortality of specialist and generalist insect pests in polycultures because of greater numbers of insect predators and parasites. Greater numbers of these natural enemies result because of the better conditions for their survival. Compared with monocultures, polycultures can provide more pollen and nectar sources (which can attract natural enemies and increase their reproductive potential), increased ground cover (which favors certain predators like carabid beetles) and increased diversity of herbivorous insects (which can serve as alternative food sources for natural enemies and make them less likely to leave when the main pest species are rare).

The second hypothesis, the resource concentration hypothesis, predicts that specialized insect pests will be less abundant in polycultures when the mixtures are composed of host and non-host crops. Specialist pests will have a more difficult time

Table 13.2 Possible effects of intercropping on insect pest populations (after Altieri and Liebman 1986)

Interference with host-seeking behavior

a) Camouflage: a host plant may be protected from insect pests by the physical presence of other overlapping plants (i.e., camouflage of bean seedlings by standing rice stubble for beanfly).

b) Crop background: certain pests prefer a crop background of a particular color and/or texture (i.e., aphids, flea beetle and Pieris rapae are more attracted to cole crops with a background of bare soil than to ones with a weedy background).

c) Masking or dilution of attractant stimuli: presence of non-host plants can mask or dilute the attractant stimuli of host plants leading to a breakdown of orientation, feeding and reproduction processes (i.e., Phyllotreta cruciferae in collards).

d) Repellent chemical stimuli: aromatic odors of certain plants can disrupt host finding behavior (i.e., grass borders repel leafhoppers in beans, populations of Plutella xylostella are repelled from cabbage/tomato intercrops).

Interference with population development and survival

a) Mechanical barriers: all companion crops may block the dispersal of herbivores across the polyculture. Restricted dispersal may also result from mixing resistant and susceptible cultivars of one crop by settling on non-host components.

b) Lack of arrestant stimuli: the presence of different host and non-host plants in a field may affect colonization of herbivores. If an herbivore descends on a non-host it may leave the plot quicker than if it descends on a host plant.

c) Microclimatic influences: in an intercropping system favorable aspects of microclimate conditions are highly fractioned, therefore insects may experience difficulty in locating and remaining in suitable microhabitats. Shade derived from denser canopies may affect feeding of certain insects and/or increase relative humidity which may favor entomophagous fungi.

d) Biotic influences: crop mixtures may enhance natural enemy complexes.

locating, remaining on and reproducing on their preferred hosts when these plants are more dispersed spatially and masked by the confusing visual and chemical stimuli presented by associated non-host crops. Some of the possible mechanisms suggested by both hypotheses that explain lower insect pest populations in multiple cropping systems are given in Table 13.2.

Evidence supports both hypotheses. Risch (1979) compared insect faunas associated with sweet potato monocultures and polycultures in Costa Rica and found more predaceous and parasitic insect individuals and fewer plant-eating insect individuals in polycultures. In experiments conducted in the Philippines, lycosid spiders preyed more heavily on maize borers when maize was grown in association with peanut than in monocultures (Litsinger and Moody 1976).

In Colombia, beans grown in dicultures with corn had 25 percent fewer leafhopper adults (Empoasca kraemeri) than monoculture beans, and population densities of the leafbeetle (Diabrotica balteata) were 45 percent lower in corn/bean plots than in bean monocultures. The incidence of the fall armyworm (Spodoptera frugiperda) was 23 percent lower in corn polycultures than in monocultures. Planting times in dicultures can also affect pest interactions. For example, further reductions in leafhopper and fall armyworm densities were achieved by establishing the companion crops 10 to 20 days before the target crop (Altieri et al. 1978).

In experiments conducted in New York state, Tahvanainen and Root (1973) observed that crucifer flea beetles were slower to colonize collard plants when collards grew in association with tomato and tobacco than when they grew in monoculture. This effect led to fewer beetle pests throughout the season. In Costa Rica, pest beetles tended to emigrate more quickly from and were less numerous in polycultures of maize, beans and squash that contained at least one non-host crop species than from monocultures of preferred hosts (Risch 1981). Similar emigration and abundance effects were noted in Michigan for striped cucumber beetles when cucumbers were grown in monocultures and in polycultures with maize and broccoli (Bach 1980). Further examinations of crop combinations that successfully prevent insect outbreaks are shown in Table 13.3.

Reductions in pest populations because of the use of polycultures can have dramatic effects on crop yields (Table 13.4). For example, competition from cowpea reduced cassava yield in polyculture when cassava whiteflies were controlled with insecticides. However, when no insecticides were used for whitefly control, yield of cassava in polyculture was higher than in monoculture. Without insecticides, the competitive effect of cowpea was more than offset by the protection it provided to the cassava (Gold, pers. comm).

Diversity and Plant Diseases

Monocultures are almost invariably prone to disease. One of the epidemiological strategies for minimizing losses from plant diseases and nematodes is to increase the species and/or genetic diversity of cropping systems. Larios (1976) documented evidence of disease buffering in various tropical intercropping schemes. Cowpea intercropped with corn showed less inoculum liberation and dissemination than in cowpea monocultures. The onset of mildew (Oidium manihotis) and scab (Sphaceloma sp.) infestation was delayed on cassava associated with beans and/or sweet potatoes. Cowpea mosaic virus and cowpea chlorotic virus occurred at lower levels in the cowpea intercropped with cassava or plantain.

The use of nonhost crops in interplantings can signficantly reduce the rate of virus spread in the field. A buffer crop such as maize, when grown between the source of peanut mottle and a susceptible soybean crop, can reduce the amount of separation required to prevent disease spread. Mosaic virus of alfalfa is more prevalent in monocultures than in mixtures with cocksfoot grass. Growing mustard or barley, which grow to heights of 66 and 41 cm respectively, together with sugarbeet stecklings lowers the incidence of beet mild yellowing virus in sugar beets (Altieri and Liebman 1986). The available examples indicate that mixtures of crop species or varieties (multilines) buffer against disease losses by delaying the onset of the disease, reducing spore dissemination or modifying microenvironmental conditions such as humidity, light, temperature and air movement (Browning and Frey 1969, Larios 1976).

Certain associated plants can function as repellants, antifeedants, growth disrupters or toxicants. In the case of soil-borne pathogens, some plant combinations and organic amendments can enhance soil fungistasis and antibiosis through indirect effects on soil organic matter (Sumner et al. 1981).

Diversity and Nematodes

A strategy based on diversity is the use of trap crops, which are host crops sown to attract nematodes but destined to be harvested or destroyed before the nematodes hatch. This tactic has been advocated for cyst nematodes, by sowing crucifers to be plowed in before the nematodes of beets can develop fully. The same objective is achieved in pineapple (Ananas comosus) plantations by planting tomatoes and destroying them before root-knot nematodes can produce eggs (Palti 1981).

There is also evidence that some plants adversely affect nematode populations through toxic action. Oostenbrink et al. (1957) showed that several varieties of Tagetes erecta and T. patula reduced the population of certain root-infecting nematode species such as Pratylenchus, Tylenchorchynchus, and Rotylenchus. The effect of marigolds on Pratylenchus eelworms appears to be due to the nematicidal action of the growing plant roots, which

Table 13.3 Selected examples of multiple cropping systems that effectively prevent insect-pest outbreaks (Altieri and Letourneau 1982)

Multiple cropping system	Pest(s) regulated	Factor(s) involved
Cabbage intercropped with white and red clover	Erioischia brassicae, cabbage aphids, and imported cabbage butterfly (Pieris rapae)	Interference with colonization and increase in ground beetles
Cotton intercropped with forage cowpea	Boll weevil (Anthonomus grandis)	Population increase of parasitic wasps (Eurytoma sp.)
Intercropping cotton with sorghum or maize	Corn earworm (Heliothis zea)	Increased abundance of predators
Cotton intercropped with okra	Podagrica sp.	Trap cropping
Strip cropping of cotton and alfalfa	Plant bugs (Lygus hesperus and L. elisus)	Prevention of emigration and synchrony in relationship between pests and natural enemies
Strip cropping of cotton and alfalfa on one side and maize and soybean on the other	Corn earworm (Heliothis zea) and cabbage looper (Trichoplusia ni)	Increased abundance of predators
Cucumbers intercropped with maize and broccoli	Acalymma vittatta	Interference with movement and tenure time on host plants
Corn intercropped with sweet potatoes	Leaf beetles (Diabrotica spp.) and leafhoppers (Agallia lingula)	Increase in parasitic wasps
Corn intercropped with beans	Leafhoppers (Empoasca kraemeri), leaf beetle (Diabrotica balteata), fall armyworm (Spodoptera frugiperda)	Increase of beneficial insects and interference with colonization
Intercropping cowpea and sorghum	Leaf beetle (Oetheca bennigseni)	Interference of air currents
Maize intercropped with canavalia	Prorachia daria and fall armyworm (Spodoptera frugiperda)	Not reported
Peaches intercropped with strawberries	Strawberry leafroller (Ancylis comptana), Oriental fruit moth (Grapholita molesta)	Population increase of parasites (Macrocentrus ancylivora, Microbracon gelechise and Lixophaga variabilis)

Table 13.3 Continued

Multiple cropping system	Pest(s) regulated	Factor(s) involved
Peanut intercropped with maize	Corn borer (Ostrinia furnacalis)	Abundance of spiders (Lycosa sp.)
Sesame intercropped with corn or sorghum	Webworms (Antigostra sp.)	Shading by the taller companion crop
Sesame intercropped with cotton	Heliothis spp.	Increase of beneficial insects and trap cropping
Squash intercropped with maize	Acalymma thiemei, Diabrotica balteata	Increased dispersion due to avoidance of host plants shaded by maize and interference with flight movements by maize stalks
Tomato and tobacco intercropped with cabbage	Flea beetles (Phyllotetra cruciferae)	Feeding inhibition by odors from non-host plants
Tomato intercropped with cabbage	Diamondback moth (Plutella xylostella)	Chemical repellency or masking

Table 13.4 Commercial root yield of cassava in monoculture and in polyculture with cowpea, with and without insecticide applications. Cassava grew at the same density in all four treatments. The experiments were conducted at Nataima, Colombia (Gold, unpub.)

Insecticides	Cassava Yield (kg/plant)	
	In Monoculture	In Polyculture
applied	1.91	1.31
not applied	0.80	1.15

xude alpha-terthienyl. In subsequent studies, Visser and Vythilingum (1959) also reported that these two marigold species considerably decreased Pratylenchus coffeae and Meloidogyne javanica populations in tea (Camellia sinensis) soil. Cultivating marigolds reduced nematodes more quickly and more effectively than keeping the tea soil fallow. There are other plants whose root extracts show nematicidal action. For example, Ambrosia spp. and Iva xanthiifolia reduce populations of P. penetrans.

Little work has been conducted on nematode suppression in intercropping systems. The nematode Anguina tritici, which enters wheat seedlings from the soil and infests the ears, has been partially controlled in India by growing Polygonum hydropiper with wheat (Triticum sp.). Also in India, the plant Sesamum orientale has been found to produce root exudates that are nematicidal to rootknot nematodes and to decrease root-knot infestation in Abelmoschus esculentus growing alongside (Altieri and Liebman 1986).

Egunjobi (1984) studied the ecology of P. brachyurus in traditional maize cropping systems of Nigeria. NPK fertilizer applications increased the numbers of the nematode more in soil under monoculture maize than in plots with maize intercropped with cowpea, groundnut or green gram.

Diversity and Weed Populations

The continuous manipulations of the fields necessary for modern crop production have favored the selection of opportunistic and competitive weeds because most weed species are stimulated by regular disturbances in monocultures. Of the factors that influence the crop/weed balance in a field, the density of crop plants and weeds plays a major role in the outcome of competition. When the cropping pattern is intensive, the level and type of weed community is a product of the crop and its management. In multiple cropping systems the nature of the crop mixtures (especially canopy closure) can keep the soil covered throughout the growing season, shading out sensitive weed species and minimizing the need for weed control. Intercropping systems of corn/mungbean and corn/sweet potato are common systems that inhibit weed competition. In these systems the complex canopies with large leaf areas intercept a significant proportion of the incident light, shading out sensitive weed species (Bantilan et al. 1974).

In general, weed suppression in intercropping systems depends on the density, relative proportions and spatial arrangement of the component crops and on the soil fertility. Mechanisms explaining overyielding and weed suppression in polycultures have been limited to resource and niche preemption, competitive exclusion and allelochemical interference (Altieri and Liebman 1986).

Allelopathy may contribute to increasing the competitiveness of crops over coexisting weeds in monocultures and polycultures. Crops such as rye, barley, wheat, tobacco and oats release toxic substances into the environment, either through root exudation or from decaying plant material, that inhibit the germination and growth of some weed species. Plant leachates from certain varieties of cucumbers have allelopathic effects on prosomillet. Root secretions from rye and oat accessions can inhibit germination and growth of weeds such as wild mustard, Brassica spp. and poppy (Papaver rhoeas) (Putnam and Duke 1978). The potential role of allelopathy in weed management is further discussed in Chapter 14.

14

Weed Ecology

Agriculture has had a major influence on the evolution of weeds. By pushing succession back to early stages, agricultural activities have maintained plant communities at immature stages. Most of the vegetational components of these communities are what modern agriculture terms "weeds." Thus far, about 250 plant species are sufficiently troublesome universally to be called weeds. Many of these weeds were introduced from distant geographic areas, or are native "opportunists" favored by particular human disturbances.

Crop monocultures seldom use all the moisture, nutrients and light available to plant growth, thereby leaving ecological niches that must be protected against invasion and competition from opportunistic weeds.

Most studies of weed ecology have emphasized the growth characteristics and adaptations that enable weeds to exploit the ecological niches left open in croplands, and the adaptive mechanisms that enable weeds to survive under conditions of maximum soil disturbance, such as under conventional tillage systems. These studies have shown that the characteristics that enable weeds to successfully colonize agroecosystems include (Baker 1974):

- o Germination requirements fulfilled in many environments: Cultivation enhances seed germination of many weed species because it augments the number of microsites (a particular location in the soil with the right germination conditions for a given species in a heterogenous environment).
- o Discontinuous germination and marked periodicity of germination: Most species germinate best at certain periods in the year. For example, Avena fatua germinates best in spring and fall, and Chenopodium album in late spring and early fall.

- Longevity of seeds: Seeds of Oenothera biennis, Verbascum blattaria and Rumex crispus can remain viable even after 80 years.
- Variable seed dormancy.
- Rapid growth through vegetative phase to flowering.
- High output of seeds under favorable conditions: For example, Amaranthus retroflexus can produce up to 110,000 seeds per plant.
- Ability to produce seeds for as long as growing conditions permit: Seed production often begins after a short period of vegetative growth.
- Self-compatible but not completely autogamous or apomictic: Many annual weeds can set seed without external pollinators.
- Adapted to cross-pollination by unspecialized visitors or wind.
- Adapted to short- and long-distance dispersal.
- Perennials have vigorous vegetative reproduction or regeneration from fragments (rhizomes, dormant buds, bulbs, taproots, etc.).
- Ability to compete interspecifically by special means (rosette, choking growth, allelochemicals).
- Ability to adapt to and tolerate variable environments.

CROP/WEED COMPETITION

Crop/weed interactions vary among different geographic regions, among different crops and even among the same crops in different situations. In fact, crop/weed interactions are overwhelmingly site-specific and season-specific. They vary according to plant species involved, density, management practices and environmental factors (Radosevich and Holt 1984). Thus, worldwide figures on crop losses may be irrelevant. However, generalizations about crop yield losses due to weeds have justified the promotion of season-long, weed-free crop systems that rely on costly chemical herbicides. This reliance has been sustained partly by chemical companies' claims that replacing herbicides with nonchemical weed control would reduce farm revenues 31 percent and result in economic losses of $13 billion (Aldrich 1984).

The end result of weed competition is a reduction in the yield or quality of the crop. In many crops, weeds left uncontrolled for the season usually prevent the production of any marketable produce. However, the outcome of this competition is affected by several factors (Zimdahl 1980):
- Period of weed growth in relation to crop emergence: Weed competition during the first third or so of the crop cycle

tends to have the greatest effect on crop yields. Generally, crop yield increases little when crops are weeded after this critical period of weed competition (Kasasian and Seeyave 1969). (Although this is a useful generalization, specific weed/crop interactions must be considered.)

o Crop type and varieties: Crops differ in their competitive ability; barley is more tolerant of interference than wheat, and wheat is more tolerant than oats. Fast-canopy-forming and tall crops with extensive leaf area suffer less from weed competition.

o Density of weed populations: Increased weed density reduces crop growth and yield.

o Weed species: Tall morning glory (Ipomoea purpurea) is more competitive in cotton than sicklepod (Cassia obtusifolia) at similar weed densities. In general, annual broadleaved weeds are more competitive than annual grass weeds at the same populations.

o Soil type: The competitive effect of weeds varies depending on soil characteristics and fertility. At high levels of fertility, little appreciable difference in crop yield occurs between weedy and weed-free crops. However, at low fertility levels weedy crops yield less than weed-free crops.

o Soil moisture: Comparative increases in yields of weedy and weed-free crops on moisture-deficient soils differ with crop and weed species. Minimal competition between soybean and Setaria spp. occurred when soil water content was either adequate or limiting during the entire season (Radosevich and Holt 1984).

o Weed physiology: The C_4 photosynthetic mechanism might have adaptive value in weeds colonizing croplands where temperatures and light intensities are high. At midday when light intensity and temperature reach peak values, C_4 weeds fix carbon dioxide at much higher rates than crops such as soybean and cotton. Weeds that have the C_4 mechanism include some grasses, Amaranthus spp. and Setaria species.

Recent evidence suggests that the presence of weeds in crop fields cannot automatically be judged damaging. Weed density/crop yield relationships are sigmoidal rather than linear. At low density, weeds do not usually affect yields, and under some circumstances certain weeds even stimulate crop growth.

For example, in the rainfed areas of the Indian arid zone, weeds such as Arnebia hispidissima, Borreria articularis and Celosia argentea increased the growth and yield of bajra (Pennisetum typhoideum) but not of til (Sesamum indicum). The presence of Indigofera cordiflora was beneficial to both crops. Similarly, in northwest India, increases in the density of the leguminous weed Triponello polycersta resulted in increased dry weight of wheat, and only at very high densities of T. polycersta (about 3,200 plants per square meter) did wheat yield decline. This positive interaction seemed to be mediated by better nitrogen nutrition of wheat due to nodular bacteria present in the roots of Triponella (Kapoor and Ramakrishnan 1975).

Studies of this nature suggest that before stressing the importance of weed control, it should be made clear whether or not a particular "weed" is harmful to a specific crop in a given area.

The degree of competition between crops and weeds can be affected by manipulating several factors. The distance between crop rows, seeding rates, specific spatial arrangements or various combinations of practices may influence the crop/weed balance (Buchanan 1977):

o Spatial arrangement of plants: Narrower crop rows result in earlier shading of the area between rows, thereby suppressing weed growth.

o Crop seeding rate: In annual cereal crops, a high seeding rate may control weeds.

o Date of planting: When crop germination coincides with emergence of the first flush of weeds, intense weed/crop interference results. One alternative is to delay planting so that yellow nutsedge can be cultivated after the first major flush of growth to reduce carbohydrates by 60 percent and subsequent vigor of the weed.

o Crop sequence: Crop rotations may influence specific weed populations. Weed seed populations in soil planted with sugarbeets were appreciably lower following a bean crop than following either barley or corn.

o Crop mixtures: Intercropping can enhance the crops' competitive abilities to suppress weeds. For example, in the Philippines, intercropping mung beans in corn reduces weed biomass and competition.

o Cover crops: Certain fall-planted cover crops can greatly reduce weed populations and biomass in the next growing season. "Tecumseh" wheat dessicated in spring or fall can reduce weed weights 76 percent and 88 percent respectively.

o Mulching: Certain plant residues provide exceptional weed control. For example, sorghum and sudangrass straw provided weed biomass reductions of 90 percent and 85 percent respectively, whereas peat moss provided only marginal reduction.

ALLELOPATHY

Competition cannot always explain suppression of plant growth in agroecosystems. At times biochemical interactions (allelopathy) occur among plants. Allelopathy is any direct or indirect harmful effect by one plant on another through the production of chemical compounds released into the environment. Contrary to competition, allelopathy occurs by the addition of a toxic factor to the environment. Allelopathy is postulated as an important mechanism by which weeds affect crop growth, and vice versa (Altieri and Doll 1978, Putnam and Duke 1978, Gliessman 1982a). Evidence strongly suggests that certain cultivars of crops such as rye, barley, wheat, tobacco and oats release toxic substances into the environment either through root exudation or from decaying plant material.

Wild types of existing crops may have possessed high allelopathic potential and this characteristic may have declined or disappeared as they were crossed and selected for other characteristics. Some accessions of Avena sp. have been shown to exhibit allelopathic influences on wild mustard (Brassica kaber). Similarly, cucumber accessions demonstrated allelopathic potential against Brassica kaber and a grass, Panicum miliaceum, under controlled environmental conditions. Under certain field conditions selected cucumber accessions inhibited the growth of prosomillet, barnyard grass and redroot pigweed. Further bioassay tests were conducted to confirm that the inhibition was due to a toxin produced by certain accessions.

Little effort has been devoted to developing crops with allelopathic potential by crossing current crop varieties with wild types. The allelopathic influence is often observed to be strongest when plants approach maturity, suggesting it could be put to best use with incipient weed problems, well after the crop has become established. Such a phenomenon would certainly be of value in influencing late-season weed control.

Allelopathy may become a viable means of weed control provided that these traits occur in wild types of cultivated species and they can be transferred to desirable cultivars. Achieving weed control by this means is inexpensive, non-polluting and requires no labels or application paraphernalia. Several alternatives exist to exploit allelopathy in agriculture:

- o Synthesize these products or their analogues for use as herbicides by isolating and identifying the toxic natural products.
- o Incorporate the toxic mechanism into cultivars through genetic manipulation.
- o Use allelopathic cover crops and mulches.
- o Manipulate weed seed behavior by using plant compounds to enhance early germination of weed seeds.

Studies comparing competitive and allelopathic components of interference between crops and weeds will establish the tolerance levels of the different weed species for each crop type.

When a crop and its accompanying weed species are considered as integral parts of the same agroecosystem, as can be observed in agroecosystems where non-crop plants are classified and managed, it becomes increasingly important to understand the complexities of the relationships between the component plant parts and the environment. Studies of the mechanisms of biotic interference between the crop and non-crop components, especially through allelopathic interactions, will become more important as the economic and ecological limitations on modern weed control practices become more restrictive. Allelopathy offers a potential alternative (Gliessman 1982a).

WEED MANAGEMENT

With changes in the relative frequencies of aggressive weed species associated with changing cropping sequences, cultivation regimes and herbicide applications, it is becoming increasingly obvious that more than one management procedure is needed to deal with the dominant weed complexes. Consequently, weed scientists have started to develop an integrated approach to weed problems, aimed at maintaining the growth of weeds at ecological, agronomic and economically acceptable levels. The approach is based on an understanding of the cultural, biological and abiotic factors causing seasonal changes in weed populations.

The central objective of weed management is to manipulate the crop/weed relationship so that growth of the crop is favored over that of the weed. Efforts have been directed at preventing weed reproduction, interrupting the recycling of weed propagules, preventing introduction of new weeds, minimizing conditions that provide niches for weed invasion and overcoming adaptations that enable weeds to persist in disturbed habitats. Cropping practices (choice of crop, rotation, crop spacing, seeding rate), tillage practices (tillage depth, minimum tillage, crop residue management) and herbicide practices are commonly used to achieve these objectives.

Any weed management program is only part of a total crop production system, so any combination of environmental manipulation, crop competition or improved cultural management techniques aimed at reducing weed levels must be compatible with other farm management schemes. In this regard, interactions between weed and pest management programs are particularly important. The farmer's soil fertility scheme is also important, as it can affect crop/weed interactions in unique ways.

Ecologists have stressed the importance of determining site-specific crop/weed relationships in terms of resource limitation, germination and growth rates. It is also important to identify the environmental interaction of weeds and the response of weeds to agroecosystem management to predict weed abundance and/or population shifts. Evidence suggests that in some circumstances manipulating one or two factors (cultivar composition, crop density, row spacing, planting date, water management, rate of applied nitrogen, tillage, crop mixture) can favorably shift the crop/weed balance. In its simplest form, weed management consists of exploiting the understanding of these relationships (Altieri and Liebman 1988).

Once the basic principles governing relationships of germination, growth and competition have been determined, management can be suggested to affect the weed community of several agroecosystems. For example, if the competitive ability of a key weed species is based on early germination, the best choice may be to plant and cultivate early. If rapid growth and canopy development are the most important strategies, times of control, weed density thresholds and fast-growing temporal and spatial crop combinations should be considered. Other relevant ecological principles on which non-chemical weed control practices can be based are shown in Table 14.1.

The Ecological Role of Weeds

The environmental simplification that characterizes modern agricultural systems has accelerated plant succession patterns in agriculture, creating specialized habitats that favor the selection of competitive and opportunistic weeds. Although weeds interfere with agricultural production, they are important biological components of agroecosystems and may be considered useful (Sagar 1974).

In many areas of Mexico, for example, local farmers do not completely clear all weeds from their cropping systems. This "relaxed" weeding is usually seen by agriculturalists as the consequence of a lack of labor and low return for the extra work. However, a closer look reveals that certain weeds are managed and even encouraged if they serve a useful purpose. In the lowland tropics of Tabasco, Mexico, there is a unique classification of non-crop plants according to use potential on one hand and effects on soil and crops on the other. Under this system, farmers recognized 21 plants in their cornfields classified as "mal monte" (bad weeds) and 20 as "buen monte (good weeds). The good weeds serve as food, medicines, ceremonial materials, teas and soil improvers (Chacon and Gliessman 1982).

Table 14.1. A list of non-chemical methods of managing weeds and the ecological principles upon which each is based (after Holt, unpublished data)

Ecological principle	Weed control practice
Reduce inputs to and increase outputs from soil seed bank	Prevention Soil solarization Weed control before seed set
Allow crop earlier space (resource) capture	Early cultivation Using crop transplants Choice of planting date
Reduce weed growth and thus space capture	Cultivation Mowing Mulching
Maximize crop growth and adaptability	Choice of crop variety Early planting
Minimize intraspecific competition of crop, maximize crop space capture	Choice of seeding rate Choice of row spacing
Maximize competitive effects of crop on weed	Planting smother or cover crops
Modify environment to render weeds less well-adapted	Rotation of crops Rotation of control methods
Maximize efficiency of resource utilization by crops	Intercropping

Similarly, the Tarahumara Indians in the Mexican Sierra depend on edible weed seedlings (Amaranthus, Chenopodium, Brassica) from April through July, a critical period before maize, bean, cucurbits and chiles mature in August through October. Weeds also serve as alternative food supplies in seasons when the maize crops are destroyed by frequent hailstorms. In a sense the Tarahumara practice a double crop system of maize and weeds that allows for two harvesets: one of weed seedlings or "quelites" early in the growing season, and another of the harvested maize late in the growing season (Bye 1981).

Weeds interact ecologically with all the other subsystems of an agroecosystem and are valuable in erosion control, conservation of soil moisture, buildup of organic matter and nitrogen in the soil and preservation of beneficial insects and wildlife (Gliessman et al. 1981). It is also a poorly recognized benefit that the soil cover provided by weeds can aid substantially in controlling erosion. One study conducted in Malawi cornfields showed that weedy groundcover reduced soil erosion losses from 12.1 tons per hectare on weeded plots to 4.5 tons per hectare on unweeded plots. An annual saving of approximately eight tons per hectare of soil should be a potentially great enough benefit to offset long-term yield reductions (Weil 1982).

Perhaps the ecological role of weeds can be best visualized by analyzing the possible consequences of a complete eradication of the weed flora from agroecosystems. Some of these include (Tripathi 1977):

o Replacement of herbicide-susceptible weed species by more resistant ones.
o Decrease in total production per unit area, because of the removal of plant biomass.
o Drastic reduction in genetic resources, since weeds contribute substantially to the existing gene pool.
o Crop plants falling victim to insects or pathogens that have so far preferred weeds.
o Reduction in the abundance of certain beneficial insects and wildlife that use weeds as alternative sources of food, shelter and breeding sites.
o Increase of erosion problems after crop harvest.
o Loss of nutrients otherwise mined and stored by weeds.

An objective analysis of the problems mentioned above should set the stage for an emphasis on weed management rather than weed control. The ecological basis of this change in emphasis has been elaborated by Bantilan et al. 1974, Buchanan and Frans 1979, Harper 1977, Sagar 1974, Tripathi 1977 and others. This re-examination of the role of weeds as ecological components can, in fact, lead to the development of guidelines for total agroecosystem management.

Weeds and the Ecology of Insect Populations

Weeds have traditionally been considered unwanted plants that reduce yields by competing with crops or by harboring insect pests and plant diseases. Between 1934 and 1963 there were 442 references relating to weeds as reservoirs of pests; 100 such references concern cereals (van Emden 1965). The Ohio Agricultural Research and Development Center recently published a series of publications concerning weeds as reservoirs for organisms affecting crops. More than 70 families of arthropods affecting crops were reported as primarily weed-associated (Bendixen and Horn 1981). More detailed examples of the role of weeds in the epidemiology of insect pests and plant diseases can be found in Thresh (1981).

Certain weeds, however, should be regarded as important components of agroecosystems that can positively affect the biology and dynamics of beneficial insects. Weeds serve as alternative sources of prey/hosts, pollen or nectar, and provide microhabitats that are not available in weed-free monocultures (van Emden 1965). The beneficial entomofauna associated with many weed species has been surveyed (Altieri and Whitcomb 1979).

In the last 20 years, research has shown that outbreaks of certain types of crop pests are more likely to occur in weed-free fields than in weed-diversified crop systems (Altieri et al. 1977). High-density crop fields with a dense weed cover usually have more predaceous arthropods than do weed-free fields. Ground beetles (Carabidae), syrphids (Syrphidae) and lady beetles (Coccinellidae) are abundant in weed-diversified systems. Relevant examples of cropping systems in which the presence of specific weeds has enhanced the biological control of particular pests are given in Table 14.2.

Work in Colombia provided experimental evidence of insect pest reduction in weed-diversified annual crops. Adult and nymph densities of Empoasca kraemeri, the main bean pest of the Latin American tropics, were reduced significantly as weed density increased in bean plots. Conversely, the chrysomelid Diabrotica balteata was more abundant in diversified bean habitats than in bean monocultures, although bean production was not affected because feeding on weeds diluted the injury to beans. In other experiments E. kraemeri populations were reduced significantly in weedy habitats, especially in bean plots with grass weeds (Eleusine indica and Laeptochloa filiformis). D. balteata densities fell by 14 percent in these systems. When grass-weed borders one meter wide surrounded bean monocultures, densities of adults and nymphs of E. kraemeri fell drastically. When bean plots were sprayed with a water homogenate of fresh grass-weed leaves, adult leafhoppers were repelled; continuous applications affected the reproduction of leafhoppers, as evinced by a reduction in the number of nymphs (Altieri et al. 1977).

Populations of insect pests and associated predaceous arthropods were sampled in simple and diversified maize habitats at two sites in north Florida during 1978 and 1979. Through various cultural manipulations, characteristic weed communities were established selectively in alternate rows with corn plots (Altieri and Letourneau 1982). Fall armyworm (Spodoptera frugiperda) incidence was consistently higher in weed-free habitats than in the corn containing natural weed complexes or selected weed associations. Corn earworm (Heliothis zea) damage was similar in all weed-free and weedy treatments, suggesting that this insect is not affected greatly by weed diversity.

The distance between plots was reduced in one site. While predators moved freely between habitats, it was difficult to identify between-treatment differences in the composition of

Table 14.2 Selected examples of cropping systems in which the presence of weeds enhanced the biological control of specific crop pests (Altieri and Letourneau 1982)

Cropping systems	Weed species	Pest(s) regulated	Factor(s) involved
Alfalfa	Natural blooming weed complex	Alfalfa caterpillar (Colias eurytheme)	Increased activity of the parasitic wasp Apanteles medicaginis
Apple	Phacelia sp. and Eryngium sp.	San Jose scale (Quadraspidiotus perniciosus) and aphids	Increased abundance and activity of parasitic wasps (Aphelinus mali and Aphytis proclia)
Apple	Natural weed complex	Tent caterpillar (Malacosoma americanum) and codling moth (Carpocapsa pomonella)	Increased activity and abundance of parasitic wasps
Beans	Goosegrass (Eleusine indica) and red sprangletop (Leptochloa filiformis)	Leafhoppers (Empoasca kraemeri)	Chemical repellency or masking
Brussels sprouts	Natural weed complex	Imported cabbage butterfly (Pieris rapae) and aphids (Brevicoryne brassicae)	Alteration of colonization background and increase of predators
Brussels sprouts	Spergula arvensis	Mamestra brassicae, Evergestis forficalis, Brevicoryne brassicae	Increase of predators and interference with colonization
Cabbage	Crataegus sp.	Diamondback moth (Plutella maculipennis)	Provision of alternate hosts for parasitic wasps (Horogenes sp.)
Collards	Ragweed (Ambrosia artemisiifolia)	Flea beetle (Phyllotreta cruciferae)	Chemical repellency or masking
Collards	Amaranthus retroflexus, Chenopodium album, Xanthium stramonium	Green peach aphid (Myzus persicae)	Increased abundance of predators (Chrysopa carnea, Coccinellidae, Syrphidae)
Corn	Giant ragweed	European corn borer (Ostrinia nubilalis)	Provision of alternate hosts for the tachinid parasite Lydella grisesens

Table 14.2 Cont.

Selected examples of cropping systems in which the presence of weeds enhanced the biological control of specific crop pests (Altieri and Letourneau 1982)

Cropping systems	Weed species	Pest(s) regulated	Factor(s) involved
Cotton	Ragweed	Boll weevil (Anthonomus grandis)	Provision of alternate hosts for the parasite Eurytoma tylodermatis
Cotton	Ragweed and Rumex crispus	Heliothis sp.	Increased populations of predators
Cruciferous crops	Quick-flowering mustards	Cabbageworms (Pieris spp.)	Increased activity of parasitic wasps (Apanteles glomeratus)
Mungbeans	Natural weed complex	Beanfly (Ophiomyia phaseoli)	Alteration of colonization background
Peach	Ragweed	Oriental fruit moth	Provision of alternate hosts for the parasite Macrocentrus ancylivorus
Soybean	Cassia obtusifolia	Nezara viridula, Anticarsia gemmatalis	Increased abundance of predators
Sugar cane	Euphorbia spp. weeds	Sugar-cane weevil (Rhabdoscelus obscurus)	Provision of nectar and pollen for the parasite Lixophaga sphenophori
Sugar cane	Borreria verticillata and Hyptis atrorubens	Cricket (Scapteriscus vicinus)	Provision of nectar for the parasite Larra americana
Sweet potatoes	Morning glory (Ipomoea asarifolia)	Argus tortoise beetle (Chelymorpha cassidea)	Provision of alternate hosts for the parasite Emersonella sp.
Vegetable crops	Wild carrot (Daucus carota)	Japanese beetle (Popillia japonica)	Increased activity of the parasitic wasp Tiphia popilliovera
Vineyards	Wild blackberry (Rubus sp.)	Grape leafhopper (Erythroneura elegantula)	Increase of alternate hosts for the parasitic wasp Anagrus epos
Vineyards	Johnson grass (Sorghum halepense)	Pacific mite (Eotetranychus willamettei)	Build-up of predaceous mites (Metaseiulus occidentalis)

predator communities. In the other site, increased distances between plots minimized such migrations, resulting in greater population densities and diversity of common foliage insect predators in the weed-manipulated corn systems than in the weed-free plots. Trophic relationships in the weedy habitats were more complex than food webs in monocultures.

Weed Management to Regulate Insect Pests

Based on the evidence discussed above, it seems that by encouraging the presence of specific weeds in crop fields, it may be possible to improve the biological control of certain insect pests (Altieri and Whitcomb 1979). Naturally, careful manipulation strategies need to be defined in order to avoid weed competition with crops and interference with certain cultural practices. In other words, economic thresholds of weed populations need to be defined, and factors affecting crop/weed balance within a crop system should be understood (Bantilan et al. 1974).

Weed management involves shifting the crop/weed balance so that crop yields are not economically reduced. It may be accomplished with herbicides, through selective cultural practices or by manipulating crops to favor crop cover rather than weeds. Suitable levels of weeds that support populations of beneficial insects can be attained within fields by designing competitive crop mixtures, allowing weed growth in alternate rows or in field margins only, use of cover crops, adoption of close row-spacings, providing weed-free periods (e.g. keeping crops free of weeds during the first third of their growth cycle from sprouting to harvest), mulching and cultivation regimes. In the state of Georgia (U.S.), populations of the velvet bean caterpillar (Anticarsia gemmatalis) and of the southern green stink bug (Nezara viridula) were greater in weed-free soybeans than in either soybeans left weedy for two or four weeks after crop emergence, or for the whole season (Altieri et al. 1981).

Changes in the species composition of weed communities are also desirable to ensure the presence of plants that affect insect dynamics. Weeded species can be manipulated by several means (Altieri and Whitcomb, 1979), such as changing levels of key chemical constituents in the soil, using herbicides that suppress certain weeds but encourage others, sowing desired weed seeds and varying weed species composition by altering the date of plowing.

15

Plant Disease Ecology and Management

Plant pathologists recently have emphasized that disease epidemics are more frequent in crops than in natural vegetation. This observation has led to the view that disease epidemics are largely the result of human interference in "the balance of nature" (Thresh 1982). The conditions that enable a pathogen to increase to epidemic levels are particularly favored by the widespread culture of genetically and horticulturally homogeneous crops, a common trend in many modern crop systems (Zadoks and Schein 1979). Extensive plantings close to major foci are particularly vulnerable, and invasion of remote sites is facilitated by the presence of intervening areas of susceptible hosts.

THE DISEASE TRIANGLE

Stated briefly, conditions necessary for the wide-scale development of a damaging disease are (Berger 1977):
1. The virulent race of the pathogen (fungi, bacteria or virus) must be present in low frequency in the host (crop).
2. The host (crop) that is susceptible to this race must be widely distributed in a region.
3. Environmental conditions must be favorable for development of the pathogen.

Together these factors form a disease triangle; their incidence and interaction result in plant disease. Increasing knowledge of the host/pathogen/environment disease triangle has enabled pathologists to apply certain ecological principles to reduce losses from epidemic disease. Although crops differ greatly in the type, permanence and stability of the habitat they provide for diseases, several features that can affect the spread of diseases in agroecosystems can be recognized (Table 15.1). In general, three epidemiological strategies can be applied to minimize losses due to disease:

Table 15.1 Some features of the crop habitat influencing the spread of crop diseases (after Thresh, 1981)

	Spread facilitated	Spread impeded
Host susceptibility	high	low
Host longevity	long	short
Host size	large	small
Vulnerable plantings	many contiguous	few scattered
Crop stands	pure	mixed
Crop spacing	close	wide
Sources of infection	many local potent	few distant less potent
Growing season	long overlapping	short distinct
Winter/dry season	mild short	extreme prolonged

1. Eliminate or reduce the initial inoculum or delay its appearance.

2. Slow the rate of increase of the pathogen.

3. Shorten the time of exposure of the crop to the pathogen by using short-season varieties or fertilization and irrigation practices that avoid slowing crop growth.

Table 15.2 summarizes the cultural, biological and chemical methods used to affect each of the three processes. The methods of biological and cultural control used up to and at the time of crop planting are the most critical for minimizing disease. Controls applied before planting include crop rotation, soil heating through solarization or burning, temporary flooding, soil amendment with large quantities of organic material and cultivation. Cultivation destroys residues and speeds up their decomposition, but also accelerates colonization by beneficial microorganisms (Cook 1986). Methods used at planting include the use of pathogen-free planting material and resistant cultivars. Genetic diversity offers great potential for genetically controlling pathogens. Genetic diversity

Table 15.2 General methods of disease control and their epidemiologic effects (after Zadoks and Schein 1979)

	x_o – amount of initial inoculum r – rate of disease increase	
	Major effect on:	
A. Avoidance of the pathogen		
1. Choice of geographic area	x_o	r
2. Choice of planting site in a local area	x_o	r
3. Choice of planting date	x_o	r
4. Use of disease-free planting stock	x_o	
5. Modification of cultural practices		r
B. Exclusion of the pathogen		
1. Treatment of seeds or planting material	x_o	
2. Inspection and certification	x_o	
3. Exclusion or restriction by plant quarantine	x_o	
4. Elimination of insect vectors	x_o	r
C. Eradication of the pathogen		
1. Biological control of plant pathogens	x_o	r
2. Crop rotation	x_o	
3. Removal and destruction of susceptible plants or diseased parts of plants		
a. Roguing	x_o	r
b. Elimination of alternate hosts and weed hosts	x_o	
c. Sanitation	x_o	
4. Heat and chemical treatments applied to planting stock	x_o	
5. Soil treatments	x_o	
D. Protection of the plant		
1. Spraying or dusting and treatment of plant propagules to protect against infection	x_o	
2. Controlling the insect vectors of pathogens		r
3. Modification of the environment		r
4. Inoculation with a benign virus to protect against a more virulent form	x_o	
5. Modification of nutrition		r
E. Development of resistant hosts		
1. Selection and breeding for resistance		
a. Vertical resistance	x_o	
b. Horizontal resistance		r
c. Two-dimensional resistance	x_o	r
d. Population resistance (multilines)		r
2. Resistance by chemotherapy		r
3. Resistance through nutrition		r
F. Therapy applied to the diseased plant		
1. Chemotherapy		r
2. Heat treatment	x_o	
3. Surgery	x_o	

can be achieved within fields by planting cultivars with different genes for resistance in different fields across an area, by planting a mixture of three or four cultivars, each having different genes for resistance, or by using cultivars having several genes for resistance within their own genetic make-up (Browning and Frey

1969).

Choosing the appropriate time and method of planting provides a means to escape pathogens. Planting either earlier or later can permit the host to pass through a vulnerable stage either before or after the pathogen produces inoculum. Variations in row spacing and depth of planting are other methods that help the crop avoid pathogen inoculum (Palti 1981).

Many of these cultural methods (crop rotation, elimination of alternative hosts, deep plowing of crop refuse, interplanting of unrelated crop types, use of barrier crops) can all be incorporated into alternative agricultural production systems; however, their adoption will greatly depend on a number of human, economic, biological and environmental factors (Table 15.3). Clearly, cultural measures have to be well adapted to the specific crop/pathogen/environmental interactions of each field, and also have to consider the demands for quick, safe and economic control of a particular disease.

Detailed treatment of epidemiologic concepts in plant disease management is found in Zadoks and Schein (1979). Palti's (1981) book provides a thorough picture of the various cultural practices for disease control.

BIOLOGICAL CONTROL OF PLANT PATHOGENS

Scientists are hopeful that biological control of plant pathogens can be exploited in modern agriculture. Principles and relevant examples are analyzed in Baker and Cook (1974). So far the most promising approach appears to be enhancing biological control agents by changing the microbial equilibrium in or around the plant to suppress the pathogen, or by introducing biological agents in the soil to suppress soil-borne plant pathogens (Papavizas 1973). The enhancement approach implies encouragement of known beneficial organisms, naturally existing in the soil, and also creation of deleterious effects on the development of pathogens. The direct approach involves mass introduction of antagonistic microorganisms in soil, with or without a food base, to inactivate pathogen propagules, thereby reducing their numbers and adversely affecting infection.

The literature on soil management practices to enhance existing microbial antagonists is voluminous. Organic amendments are recognized as initiators of two important disease-control processes: increase in dormancy of propagules and their digestion by soil microorganisms (Palti 1981). In soils amended with organic materials, propagule germination by the pathogen may not be possible, even in the presence of nutritive mixtures. Evidence suggests that the effect is relatively nonspecific in origin and

Table 15.3 Economic, social, biological and environmental factors affecting prospects for cultural control of crop diseases (Zadochs and Schein 1979)

Socio-economic factors	Prospects for cultural control	
	improve when	diminish when
Crop value and level of potential crop loss:	low	high
Cost of chemical control, relative to overall growing expenses:	high (e.g. cereals)	low
Chances for regional planning of crops to minimize inoculum build-up:	good	bad
Choices of pre-sowing practices (soil, season, topography):	numerous	few
Chances for manipulation of field conditions:	many (e.g., irrigated crops)	limited (e.g., dry farming)
Educational level of the farmer:	high	low
Pathogen factors		
Dispersal of inoculum:	splashing	wind
The wetting period needed for infection:	long	short
The rate of inoculum build-up:	rapid	slow
The temperature range for development:	narrow	wide
Susceptibility of overseasoning or dispersal of inoculum to heat and drought:	high	low
Crop host factors		
Amount of susceptible tissue available at any one time:	limited	plentiful
Range of adaptability to various growing conditions:	wide	narrow
Environmental factors		
Climatic conditions in general, in relation to optimal growth conditions:	sub-optimal, at least in some seasons	approach the optimum

involves the sum of intensified activity of the complex microbial community, including increased liberation of toxic metabolites and competition for nutrients. As microbial activity increases, the expenditure of propagule energy during dormancy presumably increases as a protection mechanism, the net result being an increase in the frequency of propagule exhaustion and death (Baker and Cook 1974).

The use of cover and legume crops, particularly green legumes plowed under, has been especially effective in biologically controlling plant pathogens. A crop of green peas or dry sorghum plowed under before planting cotton in the southwestern United States apparently provides excellent field control of phymatotrichum root rot. The effectiveness of legume cover crops for the control of take-all has been frequently demonstrated. Germinability, and possibly viability, of sclerotia of Typhula idahoensis is greatly reduced in Idaho fields where alfalfa is introduced into the rotation with wheat. Potato scab was prevented from increasing if soybeans were grown annually as a cover crop and incorporated each year before planting potatoes (Baker and Cook 1974).

Leguminous residues are rich in available nitrogen and carbon compounds, and they also supply vitamins and more complex substrates. Biological activity becomes very intense in response to amendments of this kind, and may increase fungistasis and propagule lysis. Table 15.4 provides specific examples of augmentation and disease suppression by addition of soil amendments. Table 15.5 gives examples of soil-borne fungal pathogens that can be reduced using green manure. Some examples of soil amendments that reduce nematode populations are presented in Table 15.6.

Table 15.4 Dry and decayed soil amendments that reduce some diseases caused by soil-borne fungi (after Palti 1981)

Crop and disease	Pathogen	Soil amendment
Potato wilt	Verticillium albo-atrum	Barley straw
Potato black scurf	Rhizoctonia solani	Wheat straw
Bean root rot	Thielaviopsis basicola	Oat straw, corn stover, lucerne hay
Pea root rot	Aphanomyces euteiches	Crucifer tissues
Cotton root rot	Macrophomina phaseolina	Lucerne meal, barley straw
Coriander wilt	Fusarium oxysporum f. sp. corianderi	Oil cakes
Banana wilt	F. oxysporum sp. cubense	Sugarcane residue
Avocado root rot	Phytophthora cinnamomi	Lucerne meal
Root rots of ornamentals	Phytophthora, Pythium, Thielaviopsis spp.	Composted tree bark

Table 15.5 Examples of green manures which reduce some soil-borne fungal pathogens (after Palti 1981)

Crop	Disease	Pathogen	Type of green manure	Effect on fungal population
Wheat	Take-all	Gaeumannomyces graminis	Rape, pea or mixed grasslegume	Reduction by stimulation of antagonists
	Eye-spot	Pseudocercosporella sp.	Grass	Partially reduced
Cotton	Root rot	Phymatotrichum omnivorum	Pea, Melilotus officinalis	Reduced
Potato	Scab	Streptomyces scabies	Soybean	Prevented build-up
	Black scurf	Rhizoctonia solani	Barley, oats	Slightly reduced

Table 15.6 Soil amendments found to reduce nematode populations (after Palti 1981)

Nematode species	Crop	Soil amendment tested
Meloidogyne incognita	Tomato	Sewage sludge, lucerne hay and straw, lespedeza hay, flax hay
M. javanica	Tomato, okra	Sawdust
Heterodera marioni	Peach	Crotalaria spectabilis in summer, oats in winter
H. tabacum	Eggplant	Sawdust, paper
Pratylenchus penetrans		Leaf mould + ammonium sulfate Mycelial residues from production of antibiotics Cellulose wastes from the paper industry
Hoplolaimus tylenchiformis, Xiphinema americanum		Leaf mould, sewage sludge
Helicotylenchus sp., Tylenchorhynchus sp., Meloidogyne sp.		Mustard oil cake, decayed leaves of Azidarachta indica
Pratylenchus penetrans		Oats, sudangrass
Belonolaimus longicaudatus		Activated sewage sludge
Tylenchulus semipenetrans		Castor pomace (by-product of castor oil extraction)

PART FIVE

Toward Sustainable Agriculture

16

Toward Sustainable Agriculture

Dramatic increases in crop productivity in modern agriculture have been accompanied in many instances by environmental degradation (soil erosion, pesticide pollution, salinization), social problems (elimination of the family farm; concentration of land, resources and production; growth of agribusiness and its domination over farm production; change in rural/urban migration patterns) and by excessive use of natural resources. Recently, agriculture has become increasingly subject to the constraints of inflationary petroleum prices.

THE PROBLEMS OF MODERN AGRICULTURE

The problems of modern agriculture may become even worse when conventional western technologies, developed under specific ecological and socioeconomic conditions, are applied to developing countries, as in some Green Revolution programs (see Chapter 4).

Modern farming has become highly complex, with gains in crop yield dependent on intensive management and the uninterrupted availability of supplemental energy and resources. This book is based on the premise that the modern approach is no longer appropriate in an environmentally troubled and energy-poor era; that progress toward a self-sustaining, resource-conserving, energy-efficient, economically viable and socially acceptable agriculture is warranted.

Understanding traditional farming systems may reveal important ecological clues for the development of alternative production and management systems. Industrial countries have much more to learn and probably will benefit more from the study of traditional agriculture than will developing countries. The challenge for sustainable agriculture research will be to learn how to share innovations and insights between industrial and developing countries and to end the one-way transfer of technology from the industrial world to the Third World. This exchange must be even, especially in the area of biotechnology, which depends greatly on the availability of crop genetic diversity, much of which is still preserved in traditional agroecosystems. Plant breeders from industrial countries must no longer be given free access to native germplasm in Third World countries to develop new commercial varieties that are then sold back to the Third World at considerable profit.

Realistically, the search for sustainable agricultural models will have to combine elements of both traditional and modern scientific knowledge. Complementing the use of conventional varieties and inputs with ecologically sound technologies will ensure a more affordable and sustainable agricultural production. In the U.S. and other industrial countries, adopting this approach will require major adjustments in the capital-intensive structure of agriculture. In developing countries it will also require structural changes, mainly to correct inequities in the distribution of resources, but it will also require that governments recognize rural people's knowledge as a major natural resource. The challenge will then be to maximize the use of this resource in autonomous agricultural development strategies.

When examining the problems that confront the development and adoption of sustainable agroecosystems, it is impossible to separate the biological problems of practicing "ecological" agriculture from the socioeconomic problems of inadequate credit, technology, education, political support and access to public service. Social complications, rather than technical ones, are likely to be the major barriers to any transition from high capital/energy production systems to labor-intensive, low energy-consuming agricultural systems.

A strategy to achieve sustained agricultural productivity will have to do more than simply modify traditional techniques. A successful strategy will be the outcome of novel approaches to designing agroecosystems that integrate management with the individual resource base and operate within the framework of environmental conditions (Loucks 1977). Selections will have to be based on the interaction of factors such as crop species, rotations, row spacing, soil nutrients and moisture, temperature, pests, harvesting and other agronomic procedures, and will have to accommodate the need to conserve energy and resources and protect environmental quality, public health and equitable socioeconomic development.

These systems must contribute to rural development and social equality. For this to occur, political mechanisms must encourage substitution of labor for capital, reduce levels of mechanization and farm size, diversify farm production and emphasize worker-controlled enterprises. Social reforms along these lines have the added benefits of increasing employment and reducing farmers' dependence on government, credit and industry (Levins 1973).

Obviously these proposed changes may conflict with the western capitalist view of modern agricultural development. It may be argued, for example, that increased mechanization reduces production costs or is necessary in areas where adequate labor is unavailable, and that diversified production creates problems for mechanization. Another concern is that sustainable technologies will fail to feed as many as two billion additional people by the close of this century. Each of these criticisms may be valid if analyzed within the current socioeconomic framework. But they are less valid if we recognize that sustainable agroecosystems represent profound changes that could have major social and political implications. It is here contended that most of the present and future problems of malnourishment and starvation are due more to patterns of food distribution and low access to food because of poverty, than to agricultural limits or the type of technology used in food production.

THE TRANSITION

The structure of corporate agriculture and the organization of agricultural research (which focuses on short-term problems and incremental modifications of existing technology) prevent ecological research recommendations from being incorporated into agricultural management systems (Buttell 1980a). It is obvious that agricultural enterprises will not invest in sustainable technologies for which the profits cannot be immediately captured.

In fact, emphasis on bigger yields continues, and in the 1980s this high-technology approach is epitomized by the wide-scale promotion of biotechnology, claimed as the new technological fix that can circumvent low productivity, especially in Third World agriculture (Barton and Brill 1983). It is argued that cell and tissue culture could be used immediately to accelerate the production of drought-tolerant and disease-resistant crop varieties. Embryo transplantation offers the possibility of improved livestock species. Thus, proponents contend that culturing and genetic transfer technologies can quickly provide plant materials adaptable to most areas in the world, including marginal lands.

An important dilemma for developers will be how to transfer and adapt biotechnologies to the social, economic and political conditions prevalent in developing countries. Given the present economic situation in these countries, it is reasonable to expect that biotechnologies promoted in debt-burdened developing countries might not be those best suited to local ecological and economic environments, but rather those most attractive to the large markets of the industrial nations (Kenney and Buttel 1984; Hansen et al. 1984).

As use of this technology increases, regulations will have to emerge to protect the public from environmental and health problems that may arise from the release of genetically engineered organisms (Brill 1985). Some concern exists that testing or application could lead to "ecological release" from biotic regulation of the genetically engineered organisms themselves or other biota in the same habitat. Third World bureaucracies are often slow or inefficient at enforcing safety, a situation exploited by many transnational companies to market their products banned in the developed countries.

Although biotechnology proponents argue that the plants they produce may be resistant to many pests and able to thrive in nutrient-poor soils (thus decreasing the need for pesticides and fertilizers), the approach makes farmers, especially peasants, increasingly dependent on seed companies. Given the tendency of some companies to emphasize seed/chemical "packages," farmers would also become automatically dependent on the chemicals needed to grow the seeds (Buttel 1980b). This is particularly true in the case of biotechnologies that tailor crops to specific needs (such as herbicide-resistant crops). The problem is that when farmers lose their autonomy, their production systems become governed by distant institutions over which rural communities have little control.

On the other hand, in industrial countries consideration of mixed agriculture (polycultures) is inhibited by the present land tenure system and the design of farm machinery. Therefore, research into the ecology of polycultures only makes sense as part of a broader program that includes land reform and redesign of machines (Levins 1973). Other limitations under prevailing societal conditions make the adoption of ecological farming difficult:

o Given the environmental complexity of each farming system, sustainable agricultural technologies must be site-specific. Therefore, technologies developed at experiment stations may prove inadequate in a heterogenous region of sustainable agroecosystems.

o A holistic exploration of agroecosystem design, management and structure would tend to break down disciplinary boundaries, challenging the commodity-oriented bias of current agricultural education, research and extension, and also the inflexible structure of the urban/rural markets.

o During a transitional phase, crop yields and cosmetic

quality would vary to some degree, resulting in unpredictable production, which in turn inhibits capital investment and prevents farmers from establishing stable and profitable relationships with wholesalers and processors. Many farmers will not shift to alternative systems unless there is a good prospect for monetary gain brought about by either increased output or decreased production costs. Different attitudes will depend primarily on farmers' perceptions of the short-term and long-term economic benefits of sustainable agriculture.

Apparently, it will not be possible to overcome these limitations without major changes in the structure of U.S. agriculture. The process of change could be accelerated if:

o   Agricultural research and extension focused attention on long-term problems, emphasizing small-scale, site-specific technologies developed in farmers' fields, with the active cooperation of small farmers.

o   Agricultural planning was integrated with an ecological perspective for all land use, pursuing multiple goals, such as production for food and income, improvement of nutritional quality, protection of the health of farm workers and consumers, protection of the environment, and equitable partitioning of the population between urban and rural settlement (Levins and Lewontin 1985).

o   Producer-consumer cooperatives emerged, encouraging local markets, and farmer cooperatives coordinated production goals to prevent over- or under-production, and to establish "objective" cosmetic standards.

o   Farming became a family-oriented activity, based on cooperative decisions about items such as farm management, purchase of inputs, credit and labor assignments.

o   Small farmers organized and became a strong political constituency to ensure just land reforms, appropriate legislation and improved access to public services, credit and technology.

o   Agriculture became subject to society-wide public policy decisions that subordinate agricultural resource management interests to broader political and economic interests.

o   Consumers became more effective in challenging agricultural research agendas that ignore nutrition, health and environmental issues.

The requirements to develop a sustainable agriculture clearly are not just biological or technical, but also social, economic and political, and illustrate the requirements needed to create a sustainable society. It is inconceivable to promote ecological change in the agricultural sector without advocating comparable changes in all other interrelated areas of society. The final requirement of an ecological agriculture is an evolved, conscious human being whose attitude toward nature is that of coexistence-not exploitation.

Bibliography

Abraham, C. T. and S. P. Singh. 1984. Weed management in sorghum-legume intercropping systems. J. Agric. Sci. 103: 103-115.

Adams, M. W., A. H. Ellingbae and E. C. Rossineau. 1971. Biological uniformity and disease epidemics. BioScience 21: 1067-1070.

Agboola, A. A. and A. A. Fayemi. 1972. Fixation and excretion of nitrogen by tropical legumes. Agron. J. 64: 409-412.

Aiyer, A. K. Y.N. 1949. Mixed cropping in India. Indian J. Agric. Sci. 19: 439-543.

Akobundu, I. O. 1980. Weed control strategies for multiple cropping systems of the humid and subhumid tropics. In: Weeds and Their Control in the Humid and Subhumid Tropics. Akobundu, I. O., ed. Nigeria:IITA.

Alcorn, J. B. 1984. Huastec Mayan Ethnobotany. Austin:Univ. Texas Press.

Aldrich, R. J. 1984. Weed-Crop Ecology: Principles in Weed Management. Massachusetts:Breton Publishers.

Altieri, M. A. 1983. The question of small farm development: who teaches whom? Agric. Ecosyst. Environ. 9: 401-405.

Altieri, M. A. and M. K. Anderson. 1986. An ecological basis for the development of alternative agricultural systems for small farmers in the Third World. Amer. J. Alter. Agric. 1: 30-38.

Altieri, M. A. and J. D. Doll. 1978. The potential of allelopathy as a tool for weed management in crop fields. PANS 24: 495-502.

Altieri, M. A. and J. G. Farrell. 1984. Traditional farming systems of south central Chile, with special emphasis on agroforestry. Agrofor. Syst. 2: 3-18.

Altieri, M. A. and D. K. Letourneau. 1982. Vegetation

management and biological control in agroecosystems. Crop Prot. 1: 405-430.

Altieri, M. A. and M. Z. Liebman. 1986. Insect, weed and plant disease management in multiple cropping systems. In: Multiple Cropping Systems. C. A. Francis, ed. New York: MacMillan. pp. 183-218.

Altieri, M. A. and M. Z. Liebman. 1988. Weed Management in Agroecosystems: Ecological Approaches. Florida:CRC Press.

Altieri, M.A. and L. C. Merrick. 1987. In situ conservation of crop genetic resources through maintenance of traditional farming systems. Econ. Bot. 41(1): 86-96.

Altieri, M. A. and L. L. Schmidt. 1985. Cover crop manipulation in northern California orchards and vineyards: effects on arthropod communities. Biol. Agric. Hortic. 3: 1-24.

Altieri, M. A. and W. H. Whitcomb. 1979. The potential use of weeds in the manipulation of beneficial insects. HortScience 14: 12-18.

Altieri, M. A., M. K. Anderson and L. C. Merrick. 1987. Peasant agriculture and the conservation of crop and wild plant resources. Conser. Biol. J. (in press).

Altieri, M. A., J. Davis, K. Burroughs. 1983a. Some agroecological and socioeconomic features of organic farming in California. Biol. Hort. Agric. 1: 97-107.

Altieri, M. A., D. K. Letourneau and J. R. Davis. 1983b. Developing sustainable agroecosystems. BioScience 33: 45-49.

Altieri, M. A., P. B. Martin and W. J. Lewis. 1983c. A quest for ecologically based pest management systems. Envir. Manage. 7: 91-100.

Altieri, M. A., A. van Schoonhoven and J. D. Doll. 1977. The ecological role of weeds in insect pest management systems: a review illustrated with bean (Phaseolus vulgaris) cropping systems. PANS 23: 185-205.

Altieri, M. A., C. A. Francis, A. van Schoonhoven and J. D. Doll. 1978. A review of insect prevalence in maize (Zea mays) and bean (Phaseolus vulgaris) polycultural systems. Field Crops Res. 1: 33-49.

Altieri, M. A., J. W. Todd, E. W. Hauser, M. Patterson, G. A. Buchanan and R. H. Walker. 1981. Some effects of weed management and row spacing on insect abundance in soybean fields. Prot. Ecol. 3: 334-343.

Anderson, D. T. 1981. Seeding and interculture mechanization requirements related to intercropping in India. In: Proc. Int. Workshop on Intercropping 10-13 Jan. 1979. India: ICRISAT.

Anderson, A. and S. Anderson. 1983. People and the Palm Forest. Washington:U.S. MAB Publ.

Anderson, A., A. Gely, J. Strudwick, G. Sobel and M. Pinto. 1987. Um sistema agroflorestal na varzea do Estuario Amazonico. Acta Amazonica (in press).

Andow, D. 1983. The extent of monoculture and its effects on insect pest populations with particular reference to wheat and cotton. Agric. Ecosyst. Environ. 9: 25-36.

Ardiwinata, R. O. 1957. Fish culture on paddy fields in Indonesia. Proc. Indo-Pacific Fish. Coun. 7: 119-154.

Armillas, P. 1971. Gardens on swamps. Science 174: 653-661.

Augstburger, F. 1983. Agronomic and economic potential of manure in Bolivian valleys and highlands. Agric. Ecosyst. Environ. 10: 335-346.

Azzi, G. 1956. Agricultural Ecology. London:Constable.

Bach, C. E. 1980. Effects of plant density and diversity on the population dynamics of a specialist herbivore, the striped cucumber beetle, Acalymma vittatta (Fab.). Ecology 61: 1515-1530.

Baker, H. F. 1974. The evolution of weeds. Ann. Rev. Eco. and Syst. 5: 1-24.

Baker, K. F. and R. J. Cook. 1974. Biological Control of Plant Pathogens. San Francisco:W. H. Freeman.

Bantilan, R. T., M. C. Palada and R. R. Harwood. 1974. Integrated weed management. I. Key factors affecting crop-weed balance. Philippine Weed Sci. Bull. 1: 14-36.

Bartlett, K. 1984. Agricultural Choice and Change. New York: Academic Press.

Bartlett, P. F. 1980. Adaption strategies in peasant agricultural production. Ann. Rev. Anthropol. 9: 545-573.

Barton, K. A. and W. J. Brill. 1983. Prospects in plant genetic engineering. Science 219: 671-676.

Bateson, G. 1979. Mind and Nature: a Necessary Unity. New York:Bantam.

Bayliss-Smith, T. P. 1982. The Ecology of Agricultural Systems. London:Cambridge Univ. Press.

Beets, W. C. 1982. Multiple Cropping and Tropical Farming Systems. Boulder:Westview Press.

Bendixen, L. E. and D. J. Horn. 1981. An Annotated Bibliography of Weeds as Reservoirs for Organisms Affecting Crops. III. Insects. Ohio:Agric. Res. and Dev. Center.

Beneria, L. 1984. Women in Development in Latin America. New York:Praeger.

Berger, R. D. 1977. Application of epidemiological principles to achieve plant disease control. Ann. Rev. Phytopathol. 15: 165-183.

Berlin, B., D. E. Breedlove, P. H. Raven. 1973. General principles of classification and nomenclature in folk biology. Amer.

Anthrop. 75: 214-242.
Berman, M. 1981. The Re-enchantment of the World. Ithaca: Cornell Univ. Press (republished 1984. New York:Bantam).
Bezdicek, D. F. and J. F. Powers. 1984. Organic farming: current technology and its role in a sustainable agriculture. ASA Spec. Pub. No. 46. Amer. Soc. Agron.
Blaikie, P. 1984. The Political Economy of Soil Erosion. New York:Methuen.
Blobaum, R. 1983. Barriers to conversion to organic farming practices in the midwestern United States. In: Environmentally Sound Agriculture. W. Lockeretz, ed., pp. 263-278. New York:Praeger.
Boulding, K. E. 1979. Ecodynamics: a New Theory of Societal Evolution. California:Sage Publ.
Bremen, H. and C.T. deWitt. 1983. Rangeland productivity and exploitation in the Sahel. Science 221(4618): 1341-1348.
Briggs, D. J. and F. M. Courtney. 1985. Agriculture and Environment. London:Longman.
Brill, W. J. 1985. Safety concerns and genetic engineering in agriculture. Science 227: 381-384.
Brokenshaw, D., D. Warren and O. Werner. 1979. Indigenous Knowledge Systems in Development. Washington:Univ.Press of America.
Broughton, W. J. 1977. Effects of various covers on soil fertility under Hevea brasiliensis and on growth of the tree. Agroecosystems 3: 147-170.
Browning, J. A. and K. J. Frey. 1969. Multiline cultivars as a means of disease control. Annu. Rev. Phytopathol. 7: 355-382.
Brush, S. 1977. Mountain, Field and Family. Philadelphia: Univ. Penn.
Brush, S. B. 1982. The natural and human environment of the central Andes. Mt. Res. Devel. 2: 14-38.
Brush, S. B., H. J. Carney and Z. Huaman. 1981. Dynamics of Andean potato agriculture. Econ. Bot. 35: 70-88.
Buchanan, F. A. 1977. Weed biology and competition. In: Research Methods in Weed Science, B. Truelove, ed. Alabama:Southern Weed Sci. Soc. pp. 25-41.
Buchanan, G. A. and R. E. Frans. 1979. The Role of Weeds in Agro-ecosystems. Proc. Symp. IX Int. Cong. Plant Prot. Washington, D. C. Vol. I.
Bullen, E. R. 1967. Break crops in cereal production. J. Royal Agric. Soc. England 128: 77-85.
Burdon, J. J. and R. Whitbread. 1979. Rates of increase of barley mildew in mixed stands of barley and wheat. J. Applied Ecol. 16: 253-258.

Busch, L. and W. B. Lacy. 1983. Science, Agriculture and the Politics of Research. Boulder:Westview Press.
Buschbacker, R., C. Uhl, E. Serrao. 1987. Abandoned pastures in Amazonia. II. Nutrient stocks. In: Soil and Vegetation. J. Ecol. (in press).
Buttel, F. H. 1980a. Agriculture, environment and social change: some emergent issues. In: The Rural Sociology of the Advanced Societies. F. H. Buttel and H. Newby, eds. New Jersey:Allenheld, Osmun and Co. pp. 453-488.
_____ 1980b. Agricultural structure and rural ecology: toward a political economy of rural development. Sociologia Ruralis 20: 44-62.
_____ 1986. Biotechnology and the future of agricultural research in Latin America. Seminar sobre Temas Priontarios. Cali, Colombia.
Bye, R. A. 1981. Quelites - ethnoecology of edible greens. J. Ethnobiology 1: 109-119.
Byerlee, D., M. Collinson, R. Perrin, D. Winkelman and S. Biggs. 1980. Planning Technologies Appropriate to Farmers - Concepts and Procedures. Mexico:CIMMYT.
Capra, F. 1982. The Turning Point: Science, Society, and the Rising Culture. New York:Simon and Schuster.
Carson, R. 1964. The Silent Spring. New York:Fawcett.
Centro de Educacion y Tecnologia (CET). 1983. La Huerta Campesino Organico. Chile:Inst. de Estudios y Publicaciones Juan Gynacio Molina.
Chacon, J. C. and S. R. Gliessman. 1982. Use of the "non-weed" concept in traditional tropical agroecosystems of southeastern Mexico. Agro-ecosystems 8: 1-11.
Chambers, R. 1983. Rural Development: Putting the Last First. London:Longman.
Chambers, R. and B. P. Ghildyal. 1985. Agricultural research for resource-poor farmers: the farmer first and last. Agricultural Admin. 20: 1-30.
Chang, J. H. 1968. Climate and Agriculture. Chicago:Aldine Pub. Co.
Charreau, C. and P. Vidal. 1965. Influence de l'Acacia albida Del. sur le sol, nutrition minerale et rendements des mils Pennisetum au Senegal. Agronomie Tropicale 20: 600-626.
Christanty, L., O. Abdoellah and J. Iskander. 1985. Traditional agroforestry in West Java: the pekarangan (home garden) and talun-kebun (shifting cultivation) cropping systems. In: The Human Ecology of Traditional Tropical Agriculture. G. Marten, ed. Boulder:Westview Press.
Clawson, D. L. 1985. Harvest security and intraspecific diversity in traditional tropical agriculture. Econ. Bot. 39: 56-67.

Combe, J. and G. Budowski. 1979. Classification of agroforestry techniques. In: Proc. Symp. Agroforestry Systems in Latin America. G. de las Salas, ed. Costa Rica:CATIE.

Conklin, H. C. 1956. Hananoo Agriculture. Rome:FAO.

Conklin, H. C. 1972. Folk Classification, a Topically Arranged Bibliography. New Haven:Yale Univ., Dep. Anth.

──────── 1979. An ethnoecological approach to shifting agriculture. In: Environmental and Cultural Behavior. A. P. Nayda, ed. New York:The Natural History Press.

Conway, G. R. 1981. What is an agroecosystem and why is it worthy of study? Paper presented: Workshop on Human/Agroecosystem Interactions. PESAM/EAPI. Philippines:Los Banos College.

──────── 1985. Agroecosystem analysis. Agri. Admin. 20: 1-30.

──────── 1986. Agroecosystem Analysis for Research and Development. Bangkok: Winrock Int.

Conway, R. 1981. Man versus pests. In: Theoretical Ecology. R. May, ed. Boston:Blackwell Sci.

Cook, R. J. 1986. Interrelationships of plant health and the sustainability of agriculture. Am. J. Alter. Agri. 1: 19-24.

Cook, R. J. and K. F. Baker. 1983. The Nature and Practice of Biological Control of Plant Pathogens. St. Paul:Phytopath. Soc.

Cordero, A. and R. E. McCollum. 1979. Yield potential of interplanted food crops in southeastern U. S. Agron. J. 71: 834-842.

Cox, G. W. and M. D. Atkins. 1979. Agricultural Ecology. San Francisco:W. H. Freeman and Sons.

Cromartie, W. J. 1981. The environmental control of insects using crop diversity. In: CRC Handbook of Pest Management in Agriculture. D. Pimentel, ed. Florida:CRC Press. pp. 223-250.

Crouch, L. and A. de Janvry. 1980. The class bias of agricultural growth. Food Policy 3-13.

Culliney, T. W. and D. Pimentel. 1986. Ecological effects of organic agricultural practices on insect populations. Agric. Ecosyst. Environ. 15: 253-266.

Dalton, G. E. 1975. Study of Agricultural Systems. London: Applied Sciences.

Datta, S. C. and A. K. Banerjee. 1978. Useful weeds of west Bengal rice fields. Econ. Bot. 32: 297-310.

Deere, C. D. 1982. Women and the sexual division of labor in Peru. Economic Development and Cultural Change 30: 795-811.

de Janvry, A. 1981. The Agrarian Question and Reformism in

Latin America. Baltimore:Johns Hopkins Univ. Press.
de Janvry, A. 1983. Perspectives for inter-American foundation programs in Chilean agriculture (unpublished man.).
Denevan, W. 1976. Native Populations in the Americas in 1492. Madison:Univ. Wisc. Press.
Denevan, W., J. Tracy, and J. B. Alcorn. 1984. Indigenous agroforestry in the Peruvian Amazon: Examples of Bora Indian swidden fallows. Interciencia 96: 346-357.
Dewey, K. 1981. Nutritional consequences of the transformation from subsistence to commercial agriculture. Hum. Ecol. 9 (2): 151-187.
deWit, C. T., P. G. Tow and G. L. Ennik. 1966. Competition between legumes and grasses. Versl. Land. Onder. 687: 1-30.
Doll, E. C. and L. A. Link. 1957. Influence of various legumes on the yields of succeeding corn and wheat and nitrogen content of the soils. Agron. J. 49: 307-309.
Douglass, G. K., ed. 1984. Agricultural Sustainability in a Changing World Order. Boulder:Westview Press.
Douglas, J. S. and R. A. de J. Hart. 1976. Forest Farming: Towards a Solution to Problems of World Hunger and Conservation. London:Watkins.
Doupnik, B. and Boosalis, M. G. 1980. Ecofallow—a reduced tillage system—and plant diseases. Plant Disease. 64: 31-35.
Durham, W. H. 1978. The adaptive significance of cultural behavior. Hum. Ecol. 4: 89-121.
_____ 1978. Toward a coevolutionary theory of human biology and culture. In: The Sociobiology Debate, A. L. Caplan, ed. New York:Harper and Row.
Eaglesham, A. R. J., A. Ayanaba, V. Ranga Rao and D. L. Eskew. 1981. Improving the nitrogen nutrition of maize by intercropping with cowpea. Soil Biol. and Biochem. 13: 169-171.
Edens, T. C. and D. L. Haynes. 1982. Closed system agriculture: resource constraints, management options, and design alternatives. Ann. Rev. Phytopathol. 20: 363- 395.
Edens, T. C. and H. E. Koenig. 1981. Agroecosystem management in a resource-limited world. BioScience 30: 697-701.
Egger, K. 1981. Ecofarming in the tropics - characteristics and potentialities. Plant Res. and Dev. 13: 96-106.
Egunjobi, O. A. 1984. Effects of intercropping maize with grain legumes and fertilizer treatment on populations of Pratylenchus brachyurus (Nematoda) and on the yield of maize (Zea mays). Prot. Ecol. 6: 153-167.
Ehrlich, P. 1966. The Population Bomb. New York: Ballantine.
Ellen, R. 1982. Environment, Subsistence and System. New York:Cambridge Univ. Press..

Ewel, J. J. 1986. Designing agricultural ecosystems for the humid tropics. Ann. Rev. Ecol. and Systematics 17: 245-271.
Ewel, J. J., S. Gliessman, M. Amador, F. Benedict, C. Berish, R. Bermudez, B. Brown, A. Martinez, R. Miranda and N. Price. 1984. Tropical agroecosystem structure. Agro-ecosystems 9: 183-190.
Falcon, L. A., W. R. Kane and R. S. Bethell. 1968. Preliminary evaluation of a granulosis virus for control of the codling moth. J. Econ. Entomol. 61: 1208-1213.
Farrell, J. G. 1984. The role of trees within mixed farming systems of Tlaxcala, Mexico. Master's thesis, Univ. Calif. Berkeley.
Finch, C. V. and C. W. Sharp. 1976. Cover Crops in California Orchards and Vineyards. Washington, D.C.:USDA Soil Cons. Ser..
Fleck, N. G., C. M. N. Machado and R. S. DeSouza. 1984. Eficienica da consorciacao de culturas no controle de plantas daninhas. Pesquisa Agropecuaria Brasileira 19(5): 591-598.
Francis, C. A., ed. 1986. Multiple Cropping Systems. New York: MacMillan.
Francis, C. A., C. A. Flor and S. R. Temple. 1976. Adapting varieties for intercropped systems in the tropics. In: Multiple Cropping. R. I. Papendick, P. A. Sanchez and G. B. Triplett, eds. Wisconsin:Publ. 27, pp. 235-254, Amer. Soc. Agron.
Francis, C. A. and J. H. Sanders. 1978. Economic analysis of bean and maize systems: monoculture versus associated cropping. Field Crops Res. 1: 319-335.
Gade, D. W. 1975. Plants, Man and the Land in the Vilcanota Valley of Peru. The Hague:W. Junk, Publ.
Gasto, J. C. 1980. Ecologia: el hombre y las transformacion de la naturaleza. Chile:Universitaria, Santiago.
Gasto, J. and J. M. Gasto. 1970. Uso de la tierra. El Campesino. Santiago. April, pp. 34-50.
Geertz, C. 1962. Agricultural Involution. Berkeley:Univ. Calif. Press.
Ginzburg, C. 1983. The Night Beetles: Witchcraft and Agrarian Cults in the Sixteenth and Seventeenth Centuries. London: Routledge and Kegan Paul.
Gliemeroth, G. 1950. Untersuchungen uber die einspritzung von speiserbsen. Zeit. Acker und Pflanz. 91: 519-544.
Gliessman, S. R. 1982a. Allelopathic interactions in crop/weed mixtures: applications for weed management. Paper presented: North Amer. Symp. on Allelopathy. Nov. 14-17, 1982. Univ. of Illinois, Champaign-Urbana.
Gliessman, S. R. 1982b. The agroecosystem: an integrative focus for the study of agriculture (unpubl. man.).

Gliessman, S. R., E. R. Garcia and A. M. Amador. 1981. The ecological basis for the application of traditional agricultural technology in the management of tropical agro-ecosystems. Agro-ecosystems 7: 173-185.

Glover, N. and J. Beer. 1986. Nutrient cycling in two traditional central American agroforestry systems. Agrofor. Syst. 4: 77-87.

Gomez, K. A. and A. A. Gomez. 1984. Statistical Procedures for Agricultural Research. New York:Academic Press.

Gow, D. and J. Van Sant. 1983. Beyond the rhetoric of participation. World Development 11(5): 427-446.

Graham, D. 1984. Undermining Rural Development with Cheap Credit. Boulder:Westview Press.

Grene, M. 1985. Perception, interpretation, and the sciences: toward a new philosophy of science. In: Evolultion at a Crossroads: The New Biology and the New Philosophy of Science. D. J. Depew and B. H. Weber, eds. Boston:Mass. Inst. Technol. Press.

Grigg, D. B. 1974. The Agricultural Systems of the World. London: Cambridge Univ. Press.

Grivetti, L. E. 1979. Kalahari agro-pastoral hunter-gatherers: the Tswana example. Ecol. Food and Nutrition 7: 235-256.

Hansen, M., L. Busch, J. Burkhardt, W. B. Lacy and L. R. Lacy. 1984. Plant breeding and biotechnology. BioScience 36(1):29-39.

Hardin, G. 1968. The tragedy of the commons. Science 162: 1243-1248.

Harlan, J. R. 1976. Genetic resources in wild relatives of crops. Crop Sci. 16: 329-333.

Harper, J. L. 1977. Population Biology of Plants. New York: Academic Press.

Hart, R. D. 1978. Methodologies to produce agroecosystem management plans for small farmers in tropical environments. Paper presented: Conf. on Basic Techniques in Ecological Agriculture, Third World Agric. Workshop, Int. Fed. Organic Agric. Movements, Montreal, Canada.

_____ 1979. Agroecosistemas: conceptos basicos. Costa Rica:CATIE.

Harwood, R. R. 1979. Small Farm Development - Understanding and Improving Farming Systems in the Humid Tropics. Boulder:Westview Press.

_____ 1979. The need for regional agriculture. The New Farm 1: 55-57.

_____ 1984. Organic farming research at the Rodale Research Center. In: Organic Farming: Current Technology and its Role in Sustainable Agriculture. Bezdicek, D. F.

and J. F. Powers, eds. Wisconsin:Amer. Soc. Agron.
Harwood, R. R. and E. C. Price. 1976. Multiple cropping in tropical Asia. In: Multiple Cropping, Papendick, R. I., P. A. Sanchez and G. B. Triplett, eds. Wisonsin:Amer. Soc. Agron.
Haynes, R. J. 1980. Influence of soil management practice on the orchard agroecosystem. Agro-ecosystems 6: 3-32.
Heath, M. E., R. F. Barnes and D. S. Metcalf. 1985. Forages: The Science of Grassland Agriculture. Ames:Iowa State Univ. Press.
Hecht, S. B. 1985. Environment, development and politics: capital accumulation and the livestock sector in eastern Amazonia. World Development 13(6): 663-684.
Hecht, S. B. and D. Posey. 1987. Preliminary analysis of Kayapo soil management. Human Ecol. (in press).
Hecht, S. B., A. Anderson and P. May. 1987. The subsidy from nature: palms, shifting cultivation and rural development. Human Organiz. (in press).
Heichel, G. H. 1978. Stabilizing agricultural energy needs: role of forages, rotations and nitrogen fixation. J. Soil and Water Conserv. Nov.-Dec. pp. 279-282.
Hildebrand, P. 1979. Generating technologies for traditional farmers: the Guatemalan experience. Proc. IX Int. Cong. of Plant Prot. Washington, D.C. pp. 31-34.
Hirschman, A. O. 1984. Getting Ahead Collectively: Grassroots Experiences in Latin America. New York:Pergamon Press.
Hofstetter, R. 1984. Overseeding Research Results, 1982-1984. Agron Dept. Pennsylvania:Rodale Res. Center.
Horwith, B. 1985. A role for intercropping in modern agriculture. Bioscience 35 (5): 286-291.
House, G. J. and B. R. Stinner. 1983. Arthropods in no-tillage soybean agroecosystems: community composition and ecosystem interactions. Env. Management 7: 23-28.
Huffaker, C. B. and P. S. Messenger. 1976. Theory and Practice of Biological Control. New York:Academic Press.
Huxley, P. A. 1983. Plant Research and Agroforestry. Nairobi: ICRAF.
International Crops Research Institute for the Semi-Arid Tropics (ICRISAT). 1981. Proc. Inter. Workshop on Intercropping, 10-13 Jan. 1979. Hyderabad, India.
ICRISAT. 1984. Annual Report for 1983. Patencheru, India.
Irvine, D. 1987. Successional processes. In: Indigenous Land Management Among Native Amazonians. New York:New York Botanical Garden.
Janzen, D. H. 1973. Tropical agroecosystems. Science 182: 1212-1219.

Jodha, N. S. 1981. Intercropping in traditional farming systems. In: Proc. Int. Workshop on Intercropping, 10-13 Jan. 1979. India:ICRISAT.
Johnston, H. W., S. B. Sanderson and J. A. MacLeod. 1978. Cropping mixtures of field peas and cereals in Prince Edward Island. Canad. J. Plant Sci. 58: 421-426.
Jordan, C. 1985. Nutrient Cycling in Tropical Forests. New York:John Wiley and Sons.
Kang, B. T., G. F. Wilson and T. L. Lawson. 1984. Alley Cropping: a Stable Alternative to Shifting Cultivation. Nigeria:Inter. Inst. Trop. Agric.
Kapoor, P. and P. S. Ramakrishnan. 1975. Studies on crop-legume behavior in pure and mixed stand. Agro-ecosystems 2: 61-74.
Kasasian, L. and Seeyave, J. 1969. Critical periods for weed competition. PANS 15(2): 208-212.
Kass, D. C. L. 1978. Polyculture Cropping Systems: Review and Analysis. Ithaca:Cornell Inter. Agric. Bull. No. 32.
Kenny, M. and F. H. Buttel. 1984. Biotechnology: Prospects and Dilemmas for Third World Development. Paper presented: The Right To Food. Concordia University, Montreal, Canada.
King, C. G. 1978. Ecology and Change: Rural Modernization in an African Community. New York:Academic Press.
King, F. H. 1927. Farmers of Forty Centuries. London:Cape.
Klages, K. H. W. 1928. Crop ecology and ecological crop geography in the agronomic curriculum. J. Am. Soc. Agron. 20: 336-353.
Klages, K. H. W. 1942. Ecological Crop Geography. New York: MacMillan.
Klee, G. A. 1980. World Systems of Traditional Resources Management. New York:John Wiley and Sons.
Knight, C. G. 1980. Ethnoscience and the African farmer: rationale and strategy. In: Indigenous Knowledge Systems and Development. D. Brokenshaw et al., eds. Maryland: Univ. Press of America.
Krantz, B. A. and Associates. 1974. Cropping patterns for increasing and stabilizing agricultural production in the semi-arid tropics. In: Inter. Workshop on Farming Systems. India: ICRISAT.
Krantz, B. A. 1981. Intercropping on an operational scale in an improved farming system. In: Proc. Inter. Workshop on Intercropping, 10-13 Jan. 1979. India:ICRISAT.
Kuhn, T. 1979. The Relationship Between History and History of Science. In: Interpretive Social Science. R. Rabinow and W. Sullivan, eds. Berkeley:Univ. Calif. Press.
Kurin, R. 1983. Indigenous agronomics and agricultural development in the Indus basin. Human Organiz. 42(4): 283-

294.
Lal, R. 1980. Soil erosion as a constraint to crop production. In: Priorities for Alleviating Soil-related Constraints to Food Production in the Tropics. Philippines:IRRI.

Langley, J. A., E. O. Healy and K. D. Olson. 1983. The macro implications of a complete transformation of U.S. agricultural production to organic farming practices. Agric. Ecosyst. Environ. 10: 323-334.

Larios, J. F. 1976. Epifitiologia de algunas enfermedades foliares de la yuca en diferentes sistemas de cultivo. M.S. thesis. UCR/CATIE. Turrialba, Costa Rica.

Leihner, D. 1983. Management and Evaluation of Intercropping Systems with Cassava. Colombia:CIAT.

Lentz, D. I. 1986. Ethnobotany of the Jicaque of Honduras. Econ. Bot. 40: 210-219.

Levins, R. 1973. Fundamental and applied research in agriculture. Science 181: 523- 524.

Levins, R. and M. Wilson. 1979. Ecological theory and pest management. Ann. Rev. Entomol. 25: 7-19.

Lewis, C. E., G. W. Barfon, W. G. Monson and W. C. McCormick. 1984. Integration of pines, pastures and cattle in South Georgia, U.S.A. Agrofor. Syst. 1: 277-297.

Lewis, J. K. 1959. The ecosystem concept in range management. Am. Soc. Range Manage. Abstr. 12: 23-25.

Lewis, R. and R. Lewontin. 1985. The Dialectical Biologist. Massachusetts:Harvard Univ. Press.

Liebman, M. 1988. Ecological suppression of weeds in intercropping systems: a review. In: Weed Management in Agroecosystems: Ecological Approaches. Altieri, M. A. and M. Liebman, eds. Florida:CRC Press.

Litsinger, J. A. and K. Moody. 1976. Integrated pest management in multiple cropping systems. In: Multiple Cropping. P.A. Sanchez, ed. Wisconsin:Amer. Soc. Agron. Spec. Pub. No. 27. pp. 293-316.

Lockeretz, W., R. Klepper, B. Commoner, M. Gertler, S. Fast, D. O'Leary and R. Blobaum. 1975. A Comparison of the Production, Economic Returns, and Energy Intensiveness of Corn Belt Farms that Do and Do Not Use Inorganic Fertilizers and Pesticides. St. Louis:Ctr. Biol. Nat. Syst., Wash. Univ.

Lockeretz, W., G. Shearer, R. Klepper and S. Sweeny. 1978. Field crop production on organic farms in the Midwest. J. Soil Water Cons. 33: 130-134.

Lockeretz, W., G. Shearer and D. H. Kohl. 1981. Organic farming in the corn belt. Science 211: 540-547.

Loehr, R. C. 1970. Changing practices in agriculture and their effect on the environment. CRC Crit. Rev. Env. Cont. 1:

69-99.
Lorenz, K. 1977. Behind the Mirror: A Search for a Natural History of Knowledge. New York:Harcourt Brace Jovanovich.
Loucks, O. L. 1977. Emergence of research on agro-ecosystems. Ann. Rev. Eco. Sys. 8: 173-192.
Lowrance, R., B. R. Skinner and G. S. House. 1984. Agricultural Ecosystems. New York:Wiley Interscience.
Lumsden, C. J. and E. O. Wilson. 1981. Genes, Mind, and Culture: The Coevolutionary Process. Cambridge:Harvard Univ. Press.
Lynam, J. K., J. H. Sanders and S. C. Mason. 1986. Economics and risk in multiple cropping. In: Multiple Cropping Systems. Francis, C. A., ed. New York:Macmillan.
MacDaniels, L. H. and A. S. Lieberman. 1979. Tree crops: a neglected source of food and forage from marginal lands. BioScience 29: 173-175.
Marten, G. G. 1986. Traditional Agriculture in Southeast Asia: A Human Ecology Perspective. Boulder:Westview Press.
Martin, M. P. L. D. and R. W. Snaydon. 1982. Root and shoot interactions between barley and field beans when intercropped. J. Appl. Ecol. 19: 263-272.
Matteson, P. C., M. A. Altieri and W. C. Gagne. 1984. Modification of small farmer practices for better pest management. Ann. Rev. Entomol. 29: 383-402.
McLeod, E. J. and S. L. Swezey. 1980. Survey of Weed Problems and Management Technologies. Univ. Calif. Approp. Tech. Program. Davis, Calif.:Res. Leaf. Ser.
Mead, R. and R. W. Willey. 1980. The concept of a "Land Equivalent Ratio" and advantages in yields from intercropping. Exper. Agri. 16: 217-228.
Merchant, C. 1980. The Death of Nature. Berkeley: Univ. Calif. Press.
_____ 1983. The Death of Nature: Women, Ecology and the Scientific Revolution. San Francisco:Harper and Row.
Metcalf, R. L. and W. Luckman. 1975. Introduction to Insect Pest Management. New York:Wiley-Interscience.
Midgley, J. 1986. Community Participation, Social Development and the State. New York:Methuen.
Miller, J. C. and S. M. Bell, eds. 1982. Crop Production Using Cover Crops and Sods as Living Mulches. Workshop Proceedings. Oregon State Univ., Corvallis.
Milton, J. and T. Farver. 1968. The Careless Technology. St. Louis:Wash. Univ. Press.
Minka, 1981. Issues No. 5, 6 and 7. Grupo Talpuy, Huancayo, Peru.
Monteith, L. G. 1960. Influence of plants other than the food plants of their host on finding by tachinid parasites. Can. Entomol. 92: 641-652.

Moock, J. 1986. Understanding Africa's Rural Households and Farming Systems. Boulder:Westview Press.

Moody, K. and S. V. R. Shetty. 1981. Weed management in intercrops. In: Proc. Inter. Conf. on Intercropping, 10-13 January 1979. India:ICRISAT.

Morales, H. L. 1984. Chinampas and integrated farms: learning from the rural traditional experience. In: Ecology in Practice: Vol. 1. F. diCastri, F. W. G. Baker and M. Hadley, eds. Dublin:Ecosystem Management, pp. 188-195.

Moreno, R. A. 1975. Diseminaction de Ascochyta phaseolorum en variedades de frijol de costa bajo diferentes sistemas de cultivo. Turrialba 25(4): 361-364.

Moreno, R. A. 1979. Crop protection implications of cassavaintercropping. In: Intercropping with Cassava: Proc. of the Int. Workshop, Trivandrum, India. 27 Nov-1 Dec 1978. E. Weber, B. Nestal and M. Campbell, eds. Canada: Int. Devel. Res. Centre.

Mueller, D. H., T. C. Daniel and R. C. Wendt. 1981. Conservation tillage: best management practice for nonpoint runoff. Env. Manage. 5: 33-53.

Murdoch, W. W. 1975. Diversity, complexity, stability and pest control. J. Appl. Ecol. 12(3): 795-807.

Naban, G. P. 1983. Papago Indian Fields: Arid Lands Ethnobotany and Agricultural Ecology. Unpublished Ph.D. Diss. Univ. Arizona, Tucson.

Nair, P. K. R. 1979. Intensive multiple cropping with coconuts in India. Principles, programmes and prospects. Advances in Agronomy and Crop Science No. 6. Berlin:Verlag.

____ 1983. Tree integration on farmlands for sustained productivity of small holdings. In: Environmentally Sound Agriculture. W. Lockeretz, ed. New York:Praeger, pp. 333-350.

____ 1984. Soil Productivity Aspects of Agroforestry. Nairobi:ICRAF.

Natarajan, M. and R. W. Willey. 1980. Sorghum-pigeonpea intercropping and the effects of plant population density. 2: Resource use. J. Agri. Sci. 95: 59-65.

____ 1981. Growth studies in sorghum/pigeonpea intercropping with particular emphasis on canopy development and light interception. In: Proc. Int. Workshop on Intercropping, 10-13 Jan. 1979. India: ICRISAT.

____ 1986. The effects of water stress on yield advantages of intercropping systems. Field Crops Res. 13: 117-131.

National Academy of Sciences (NAS). 1977. Leucaena: Promising Forage and Tree Crop for the Tropics. Washington, D.C.: NAS.

Netting, R. Mc. 1974. Cultural Ecology. California:Cummings.
_____ 1974. Agrarian ecology. Ann. Rev. Anth. 1: 21-55.
Norgaard, R. B. 1981. Sociosystem and ecosystem coevolution in the Amazon. J. Env. Econ. Manage. 8: 238-254.
_____ 1984. Coevolutionary agricultural development. Econ. Dev. and Cult. Change 60: 160-173.
Norman, D. W. 1977. The rationalization of intercroppoing. African Envir. 2(4)/3(1): 97-109.
_____ 1979. The farming systems approach: relevancy for small farmers. In: Increasing the Productivity of Small Farms. H.S. Karaspan, ed. Lahore:Pakistan Centro. pp. 37-49.
Norman, M.J.T. 1979. Annual Cropping Systems in the Tropics: An Introduction. Gainesville:University Presses of Florida.
Nye, P.H. and D.J. Greenland. 1961. The Soil Under Shifting Cultivation. England:Comm. Agri. Bur.
O'Brien, T. A., J. Moorley and W. J. Whittington. 1967. The effect of management and competition on the uptake of 32P by ryegrass, meadow fescue and their natural hybrid. J. Appl. Ecol. 4: 513-520.
Odum, E. 1984. Properties of agro-ecosystems. In: Agricultural Ecosystems. Lowrance et al., eds. New York:Wiley Intersci.
Oelhaf, R. C. 1978. Organic Agriculture. New Jersey:Allanheld, Osmun and Co. Pub. Inc.
Okigbo, B. N. and D. J. Greenland. 1976. Intercropping systems in tropical Africa. In: Multiple Cropping. R. I. Papendick, P. A. Sanchez and G. B. Triplett, eds. Wisconsin:Amer. Soc. Agron..
Oldeman, R.A.A. 1981. The design of ecologically sound agroforests. In: Viewpoints on Agroforestry. K. F. Wiersum, ed. The Netherlands:Ag. Univ. Wageningen.
Oostenbrink, M., K. Kuiper and J. J. S'Jacob. 1957. Tagetes als feindpflanzen von Pratylenchus Arten. Nematologica 2, Suppl: 424-433.
Osiru, D. S. O. and R. W. Willey. 1972. Studies on mixtures of dwarf sorghum and beans with particular reference to plant population. J. Agri. Sci. 79: 531-540.
Palada, M. C., S. Ganser, R. Hofstetter, B. Volak and M. Culik. 1983. Association of interseeded legume cover crops and annual row crops in year-round cropping systems. In: Environmentally Sound Agriculture. W. Lockeretz, ed. New York:Praeger.
Palmer, R. and N. Parsons. 1977. The Roots of Rural Poverty in Central and Southern Africa. Berkeley:Univ. Calif. Press.
Palti, J. 1981. Cultural Practices and Infectious Crop Diseases. New York:Springer-Verlag.

Papadakis, J. 1938. Compendium of Crop Ecology. Buenos Aires, Argentina:Papadakis.
_____ 1941. Small grains and winter legumes grown mixed for grain production. J. Amer. Soc. Agron. 33: 504-511.
Papavizas, G. C. 1973. Status of applied biological control of soil-borne plant pathogens. Soil Biol. Biochem. 5: 709-720.
Papendick, R. I., P. A. Sanchez and G. B. Triplett. 1976. Multiple Cropping. Wisconsin:ASA Spec. Pub. No.27.
Papendick, R. I., L. F. Elliott and R. G. Dohlgren. 1986. Environmental consequences of modern production agriculture. Amer. J. of Altern. Agric. 1: 3-10.
Parr, J. F., Papandick, R. I. and I. G. Youngberg. 1983. Organic farming in the United States: principles and perspectives. Agro-ecosystems 8: 183-201.
Pearce, A. 1975. The Latin American Peasant. London:Frank Cass.
Pearce, A. 1980. Seeds of Plenty; Seeds of Want: Social and Economic Implications of the Green Revolution. New York:Oxford Press.
Perelman, M. 1977. Farming for Profit in a Hungry World. New Jersey:Allanheld, Osmun and Co. Pub. Inc.
Perrin, R. M. 1977. Pest management in multiple cropping systems. Agro-ecosystems 3: 93-118.
Perrin, R. K., D. L. Winkelmann, E. R. Moscardi and J. R. Anderson. 1979. From Agronomic Data to Farmer Recommendations: an Economic Training Manual. Info. Bull. 27. Mexico City:CIMMYT.
Phillips, R.E., R.L. Blevins, G.W. Thomas, W.W. Frye and S. H. Phillips. 1980. No-tillage agriculture. Science 208: 1108-1113.
Phillips, S. H. and H. M. Young, Jr. 1973. No-tillage Farming. Wisconsin:Reiman Assoc., Inc.
Pimentel, D. 1973. Extent of pesticide use, food supply and pollution. Proc. N.Y. Ent. Soc. 81: 13-33.
Pimentel, D. and M. Pimentel. 1979. Food, Energy and Society. London:Edward Arnold.
Pimentel, D., G. Berardi and S. Fast. 1983. Energy efficiency of farming systems: organic and conventional agriculture. Agric. Ecosyst. Environ. 9: 359-372.
Pimentel, D. and N. Goodman. 1978. Ecological basis for the management of insect populations. Oikos 30: 422-437.
Posey, D. 1984. Ethnoecology as applied anthropology in Amazonian development. Human Organiz. 43(2): 95-107.
_____ 1985. Indigenous management of tropical forest ecosystems: the case of the Kayapo Indians. Agrofor. Syst. 3:2.

Power, J. F. and J. W. Doran. 1984. Nitrogen use in organic farming. In: Nitrogen in Crop Production. Wisconsin: ASA-CSSA-SSSA, pp. 585-592.
Price, P. W. and G. P. Waldbauer. 1975. Ecological aspects of pest management. In: Introduction to Insect Pest Management. R.L Metcalf and W.H. Luckman, eds. New York:Wiley-Interscience, pp. 37-73.
Protheroe, R. M. 1972. People and Land in Africa South of the Sahara. London:Oxford Univ. Press.
Pullin, S. V. and Z. H. Shehadeh. 1980. Integrated Agriculture-Agriculture Farming Systems. Manila:Int. Center for Living Aquatic Res. Manage..
Putnam, A. R. and J. DeFrank. 1983. Use of phytotoxic plant residues for selective weed control. Crop Prot. 2: 173-181.
Putnam, A. R. and W. B. Duke. 1978. Allelopathy in agroecosystems. Annu. Rev. Phytopathol. 16: 431-451.
Quinne, J.F. and A.E. Dunham. 1983. On hypothesis testing in ecology and evolution. Amer. Nat. 122(5): 22-37.
Rabb, R. L. 1978. A sharp focus on insect populations and pest management from a wide-area view. Bull. Entomol. Soc. Amer. 24(1): 55-61.
Radke, J. K. and R. T. Hagstrom. 1976. Strip intercropping for wind protection. In: Multiple Cropping. R. I. Papendick, P. A. Sanchez and G. B. Triplett, eds. Wisconsin:Amer. Soc. Agron.
Radosevich, S. R. and J. S. Holt. 1984. Weed Ecology: Implications for Vegetation Management. New York:John Wiley and Sons.
Raeburn, J. R. 1984. Agriculture: Foundations, Principles and Development. New York:John Wiley and Sons.
Rao, M. R. and R. W. Willey. 1980. Evaluation of yield stability in intercropping: studies on sorghum/pigeonpea. Exper. Agri. 16: 105-116.
Reddy, M. S. and R. W. Willey. 1981. Growth and resource use studies in an intercrop of pearl millet/groundnut. Field Crops Res. 4: 13-24.
Rhodes, R.. and R. Booth. 1982. Farmer back to farmer: a model for generating agricultural technology. Agri. Admin. 11(2): 127-137.
Richards, A. 1939. Land Labor and Diet in Northern Rhodesia. London:Routledge and Kegan Paul.
Richards, P. 1984. Indigenous Agricultural Revolution. Boulder:Westview Press.
_____ 1986. Coping with Hunger: Hazard and Experiment in African Rice Farming. Boulder:Westview Press.
Risch, S. J. 1979. A comparison by sweep sampling of the insect

fauna from corn and sweet potato monocultures and dicultures in Costa Rica. Oecologia 42: 195-211.

_____ 1981. Insect herbivore abundance in tropical monocultures and polycultures: an experimental test of two hypotheses. Ecology 62: 1325-1340.

Risch, S. J. 1983. Intercropping as cultural pest control: prospects and limitations. Envir. Manage. 7(1): 9-14.

Risch, S. J., D. Andow and M. A. Altieri. 1983. Agroecosystem diversity and pest control: data, tentative conclusions and new research directions. Env. Entomol. 12: 625-629.

Root, R. B. 1973. Organization of a plant-arthropod association in simple and diverse habitats: the fauna of collards (Brassica oleracea) Ecol. Monogr. 43: 95-124.

Ruthenberg, H. 1971. Farming Systems of the Tropics. London:Oxford Univ. Press.

Sagar, G. R. 1974. On the ecology of weed control. In: Biology in Pest and Disease Control. D. Price Jones and M. E. Solomon, eds. England:The 13th Symp. Brit. Ecol. Soc.

Sanders, J. H. and D. V. Johnson. 1982. Selecting and evaluating new technology for small farmers in the Colombian Andes. Mt. Res. Dev. 2(3): 307-316.

Schumacher, R. 1973. Small is Beautiful. London:Abacus.

Scott, J. 1978. The Moral Economy of the Peasant. Madison: Univ. Wiscon.

_____ 1986. Weapons of the Weak: Everyday Forms of Peasant Resistence. New Haven:Yale Press.

Sen, A. 1981. Poverty and Famines. New York:Oxford Press.

Senanayake, R. 1984. The ecological, energetic and agronomic systems of ancient and modern Sri Lanka. In: Agricultural Sustainability in a Changing World Order. Douglass, G. K., ed. Colorado:Westview Press, pp. 227-238.

Shaner, W. W., P. F. Philipp and W. R. Schmehl. 1982. Farming Systems Research and Development: Guidelines for Developing Countries. Boulder:Westview Press.

Shenk, M. D. and J. L. Saunders. 1983. Insect population responses to vegetation management systems in tropical maize production. In: No-tillage Crop Production in the Tropics. I.O. Akobundo and A. E. Deutsch, eds. Oregon: Int. Plant Protection Center, pp. 73-85.

Simon, J. and H. Kahn. 1985. The Resourceful Earth. New York:Oxford Press.

Shetty, S. V. R. and A. N. Rao. 1981. Weed management studies in sorghum/pigeonpea and pearl millet/groundnut intercrop systems—some observations. In: Proc. Int. Workshop on Intercropping 10-13 Jan. 1979. India:ICRISAT, pp. 238-248.

Smith, J. R. 1953. Tree Crops: A Permanent Agriculture. New

York:Devin-Adair.
Sober, Elliot, ed. 1984. Conceptual Issues in Evolutionary Biology: An Anthology. Cambridge:Mass. Inst. Techn. Press.
Soria, J., R. Bazan, A. M. Pinchinat, G. Paez, N. Mateo, R. Moreno, J. Fargas and W. Forsythe. 1975. Investigacion sobre sistemas de produccion agricola para el pequeno agricultor del tropico. Turrialba 25(3): 283-293.
Southwood, T. R. E. and M. J. Way. 1970. Ecological background to pest management. In: Concepts of Pest Management. R. L. Rabb and F. E. Guthrie, eds. North Carolina:North Carolina State University, pp. 6-28
Spedding, C. R. W. 1975. The Biology of Agricultural Systems. London:Academic Press.
Sprague, M. A. and G. B. Triplett, eds. 1986. No-tillage and Surface Tillage Agriculture. New York:John Wiley and Sons.
Steiner, K. G. 1984. Intercropping in Tropical Smallholder Agriculture with Special Reference to West Africa, 2nd ed. Deutsche Gesell. Techn. Zusam. (GTZ). Federal. Rep. Germany:Eschborn.
Steward, J. H. 1977. Evolution and Ecology: Essays on Social Transformation. Urbana:Univ.Illinois Press.
Sumner, D. R. 1982. Crop rotation and plant productivity. In: CRC Handbook of Agricultural Productivity, vol. I. M. Rechcigl, ed. Florida:CRC Press, pp. 273-313.
Sumner, D. R., B. Doupnik, Jr. and M. G. Boosalis. 1981. Effects of reduced tillage and multiple cropping on plant diseases. Annu. Rev. Phytopathol. 19: 167-187.
Suryatna, E. 1979. Cassava intercropping patterns and management practices in Indonesia. In: Intercropping with Cassava: Proc. Int. Workshop.Trivandrum,India, 27 Nov.-1 Dec. 1978. E. Weber, B. Nestel and M. Campbell, eds. Canada:Int. Dev. Res. Centre, pp. 35-36
The Ecologist. 1972. Blue print for survival. The Ecologist 2: 1-43.
Thorne, D. W. and M. D. Thorne. 1979. Soil, Water and Crop Production. Westport:AVI Pub. Co.
Thresh, J. M. 1981. Pests, Pathogens and Vegetation: The Role of Weeds and Wild Plants in the Ecology of Crop Pests and Diseases. Massachusetts:Pitman Pub., Inc.
_____ 1982. Cropping practices and virus spread. Annu. Rev. Phytopathol. 20: 193-218.
Tischler, W. 1965. Agrookologie. Jene:Eustan Fisher.
Todd, R. L., R. Leonard and L. Asmussed, eds. 1984. Nutrient Cycling in Agricultural Ecosystems. Michigan:Ann Arbor Sci. Publ.
Toledo, V. M. 1980. La ecologia del modo campesino de

produccion. Anthropologia y Marxismo 3: 35-55.
Toledo, V. M., J. Cararbias, C. Mapes, and C. Toledo. 1985. Ecologia y autosuficiencia alimentaria. Mexico City:Siglo Veintiuno Ed..
Toulmin, Stephen. 1982. The Return of Cosmology: Postmodern Science and the Theology of Nature. Berkeley:Univ. Calif.
Trenbath B. R. 1976. Plant interactions in mixed crop communities. In: Multiple Cropping. R. I. Papendick, P. A. Sanchez and G. B. Triplett, eds. Wisconsin:Amer. Soc. Agron. pp. 129-170.
Trenbath, B. R. 1983. The dynamic properties of mixed crops. In: Frontiers of Research in Agriculture. S. K. Roy, ed. India:Indian Stat. Inst. pp. 265-286.
Tripathi, R. S. 1977. Weed problems - an ecological perspective. Trop. Ecol. 18: 138-148.
Tripathi, B. and C. M. Singh. 1983. Weed and fertility management using maize/soyabean intercropping in the northwestern Himalayas. Trop. Pest Manage. 29(3): 267-270.
Troeh, F. R., J. A. Hobbs and R. L. Donahue. 1980. Soil and Water Conservation for Productivity and Environmental Protection. New York:Prentice Hall.
Tustin, J. R., R. L. Knowles and B. K. Klomp. 1979. Forest farming: a multiple land-use production system in New Zealand. For. Ecol. Manage. 2: 169-189.
Uhl, C. 1983. Nutrient uptake and nutrient-retention related characteristics of common successional species in the upper Rio Negro region of the Amazon basin. (unpublished man.).
Uhl, C. and P. Murphy. 1981. A comparison of productivities and energy values between slash and burn agriculture and secondary succession in the upper Rio Negro region of the Amazon Basin. Agro-ecosystems 7: 63-83.
Uhl, C. and C. Jordan. 1984. Vegetation and nutrient dynamics during the first five years of succession following forest cutting and burning in the Rio Negro region of Amazonia. Ecology 65: 1476-1490.
Uhl, C., R. Buschbacker and A. Serrao. 1987. Abandoned pastures in Amazonia. I. Patterns of plant succesion. J. Ecol. (in press).
USDA. 1973. Monoculture in Agriculture: Extent, Causes, and Problems - Report of the Task Force on Spatial Heterogeneity in Agricultural Landscapes and Enterprises. Wash., D.C.:USDA.
USDA. 1980. Report and Recommendations on Organic Farming. Wash., D.C.:USDA.
van den Bosch, R. and A. D. Telford. 1964. Environmental modification and biological control. In: Biological Control

of Insect Pests and Weeds. P. DeBach, ed. New York: Reinhold, pp. 459-488.

van Dyne, G. M. 1969. The Ecosystem Concept in Natural Resource Management. New York:Academic Press.

van Emden, H. F. 1965. The role of uncultivated land in the biology of crop pests and beneficial insects. Sci. Hort. 17: 121-136.

van Emden, H. F. and G. F. Williams. 1974. Insect stability and diversity in agroecosystems. Annu. Rev. Entomol. 19: 455-475.

Vandermeer, J. 1981. The interference production principle: an ecological theory for agriculture. BioScience. 31: 361-364.

Vink, A. P. A. 1975. Land Use in Advancing Agriculture. New York:Springer-Verlag.

Visser, T. and M. K. Vythilingam. 1959. The effect of marigolds and some other crops on the Pratylenchus and Meloidogyne populations in tea soil. Tea Quart. 30: 30-38.

Wade, M. K. and P. A. Sanchez. 1984. Productive potential of an annual intercropping scheme in the Amazon. Field Crops Res. 9: 253-263.

Wasserstrom, R. 1982. Land and Labor in Chiapas. Berkeley:Univ. Calif. Press.

Watts, M. 1983. Silent Violence. Berkeley:Univ. Calif. Press.

Webster, C. C. and P. N. Wilson. 1980. Agriculture in the Tropics. Trop. Agri. Series. London:Longman, pp. 9-12.

Weil, R. N. 1982. Maize-weed competition and soil erosion in unweeded maize. Trop. Agri. 59: 207-213.

Whittington, W. J. and T. A. O'Brien. 1968. A comparison of yields from plots sown with a single speices or a mixture of grass species. J. Appl. Ecol. 5: 209-213.

Whittlesay, D. 1936. Major agricultural regions of the earth. Ann. Assoc. Amer. Geog. 26: 199.

Wiersum, K. F., ed. 1981. Viewpoints on Agroforestry. The Netherlands:Agricultural University, Wageningen.

Wilken, G. C. 1969. The ecology of gathering in a Mexican farming region. Econ. Bot. 24: 286-295.

_____ 1977. Integrating forest and small-scale farm systems in middle America. Agro-ecosystems 3: 291-302.

Wilkes, H. G. 1977. Hybridization of maize and teosinte in Mexico and Guatemala and the improvement of maize. Econ. Bot. 31: 254-293.

Willey, R. W. and D. S. O. Osiru. 1972. Studies on mixtures of maize and beans with particular reference to plant population. J. Agri. Sci. 79: 519-529.

Wilsie, C. P. 1962. Crop Adaptation and Distribution. San Francisco:W.H. Freeman Co..

William, R. D. 1981. Complementary interactions between weeds, weed control practices, and pests in horticultural cropping systems. HortScience 16: 508-513.

Williams, D. E. 1985. Tres arvenses Solanaceas comestibles y su proceso de domesticacion en Tlaxcala, Mexico. Master's thesis, Colegio de Postgraduados, Chapingo, Mexico.

Wilson, G. F. and B. T. Kang. 1981. Developing stable and productive biological cropping systems for the humid tropics. In: Biological Husbandry: a Scientific Approach to Organic Farming. B. Stonehouse, ed. London:Butterworths, pp. 193-203.

Wittwer, S. H. 1975. Food production: technology and the resource base. Science 188: 579-588.

Wolf, E. 1982. Europe and the People Without History. Berkeley: Univ. Calif .Press.

Woodmansee, R. 1984. Comparative nutrient cycles of natural and agricultural ecosystems. In: Agricultural Ecosystems. R. Lowrance, B. R. Skinner, and G. S. House, eds. New York:John Wiley and Sons. pp. 145-157.

Zadoks, J. C. and R.D. Schein. 1979. Epidemiology and Plant Disease Management. New York:Oxford Univ. Press.

Zandstra, H. G., E. C. Price, J. A. Litsinger and R. A. Morris. 1981. A Methodology for On-farm Cropping Systems Research. Philippines:IRRI.

Zimbdahl, R. L. 1980. Weed-Crop Competition: a Review. Corvallis:Int. Plant Prot. Center, Oregon State Univ.

About the Contributors

MIGUEL A. ALTIERI is associate professor and entomologist at the Division of Biological Control, University of California, Berkeley. He teaches courses in agricultural ecology, rural development and integrated pest management in developing countries, agroforestry systems, insect ecology and biological control, and weed science. He also conducts international short courses in the United States and Latin America on biological control and agroecology in addition to researching methods to enhance naturally occurring and introduced biological control agents of pests in annual agricultural systems and orchards. He is currently surveying sustainable agriculture projects in Latin America.

JOHN G. FARRELL has worked on several domestic and international projects in the areas of plant ecology and agroforestry and has collaborated in the survey of traditional farming systems. His pioneer master's thesis "The Role of Trees Within Mixed Farming Systems of Tlaxcala, Mexico" was completed at the University of California, Berkeley in 1984. He has been a guest lecturer in agroforestry at the University of Hawaii and is now Field Research Manager at the Student Farm for the University of California, Santa Cruz.

SUSANNA B. HECHT is visiting assistant professor in the Graduate School of Architecture and Urban Planning at the University of California, Los Angeles where she teaches courses in agriculture and ecology, regional development and policy, resource science, and soil science. She also consults with government and non-governmental groups on topics related to land use and agricultural policy in the Amazon and humid tropics. Countries where she has consulted include Colombia, Peru, Mexico, Brazil and the Dominican Republic.

MATT LIEBMAN received his Ph.d. in 1986 on studies of resource use, crop yield and weed suppression in intercrop/weed mixtures. He is presently assistant professor of sustainable agriculture in the Department of Plant and Soil Science at the University of Maine where he is coordinator for a four year sustainable agriculture curriculum. His research focuses on minimizing inputs of nonrenewable resources into agricultural production while protecting soil productivity.

RICHARD B. NORGAARD is associate professor in the Energy and Resource Program and the Department of Agricultural and Resource Economics at the University of California, Berkeley. He teaches a course in resource ecology and alternative development strategies and does consulting work with the World Bank, the United Nations Food and Agriculture Organization (UNFAO), The United States Agency for International Development (USAID) and other agencies on environment and development.

Index

Agricultural ecology,
methodological approaches, 9
Agricultural science, 6
Agricultural system,
design of, 59
Agriculture, Andean, 82
Agroecological world view, 23
Agroecology, 1, 4, 21, 27, 159
Agroecosystem design, 62
Agroecosystems, 29
    characteristics of, 31
    classification of, 29
Agroforestry, 149
    characteristics of, 149
    examples of, 157
Agroforestry systems, 149
    classification of, 150
    design of, 155
    role of trees in, 150
Artificial ecosystem, 39
Capital, 56
Capital resources, 33
Chinampas, 87
Conventional farming, 39
    energy use in, 108
Crop diversity, 73
Crop rotation, 139
Crop yield, 48, 107, 146
Cropping systems, 64, 109, 142
    cover cropping, 127
    crop rotation,
      benefits of, 139
    living mulch, 132
      with vegetables, 136
    minimum tillage, 142
      effects on soil and plant growth, 143
      effects on pests, 144
      energy requirements, 146
    monoculture, 163
    non-tillage, 129, 142
    polyculture, 115, 164
      effects on diseases, 121
      effects on insects, 121
      effects on nematodes, 122
      effects on weeds, 122
      resource use, 119
      yield advantages, 116
      yield stability, 118
    sod strip intercropping, 135
    tillage, 130
    with legumes, 133
      overseeding, 133
      sod-based rotations, 134
Development studies, 17
Diversity, 84, 161
    and insect populations, 164
    and plant diseases, 166
    and nematodes, 167
    and weed populations, 170
    spatial and temporal, 71

Ecological processes in
the agroecosystem, 33
Ecological view, 5
Ecology, 13
Ecosystem, 24
Environmentalism, 10
Epistemology, 3, 21
Equity, 42
Farming systems research, 50
Field procedures, 53
Food system, 29
Green Revolution, 17-18
Human resources, 32
Hydrological processes, 36, 154
Inappropriate technology, 47-48
Indigenous production
systems, 14
Inputs, 34
Integrated pest management, 161
Land equivalent ratio, 73, 116
Land use capability, 62
Legumes (see cropping systems)
Legume sod-based rotations
(see cropping systems)
Living mulches, 132
Minimum tillage systems, 142
Modern agriculture, problems
of, 108, 195
Multiple cropping (see cropping
systems)
Natural resources, 31
Nematodes, 122 (see
polycultures)
Nkomanjila system, 80
Nkule system, 81
Nutrients, cycling of, 13
Organic farming, 107
    and wildlife, 112
    characteristics of, 107
    constraints to, 113
    cultural practices, 109
    insecticide use, 110
    labor requirements, 113
    plant nutrients and
    soil organic matter, 111
    weed control, 109

Outputs, 34
Paddy rice culture, Southeast
Asia, 75
Pekarangan (Javanese
home-garden), 77
Pest management, 9, 39, 159
    agroecology and, 159
    integrated pest
    management, 161
Plant diseases, 187
    cultural control of, 188
    methods of biological
    control of, 190
    the disease triangle, 187
Polycultures, advantages
of, 73
Production, primary, 62
Productivity, 44, 57
Rainfed agriculture, 37
Shifting cultivation, 78
Soil moisture, 36
Stability, 42
Succession, 63
    control of, 71
Sustainable agriculture, 195
    transition to, 197
Sustainability, 41
    elements of, 60
Talun-keban, Java, 77
Third World development, 17
Tillage, 142
Traditional agriculture, 69
    ecological features of, 69
Java, 76
Traditional
agroecosystems, 69
    and genetic resources, 90
Traditional and modern
technologies, 93
    alley cropping in
    Africa, 94
    Andean agriculture,
    Bolivia, 95
    aquaculture systems,
    Veracruz, Mexico, 103

Minka project, Peru, 96
modular system, Tabasco,
Mexico, 101
Traditional farming
systems, 69
    Andean agriculture, 83, 95
    Mediterranean Chile, 84, 96
    Javanese traditional
    agriculture, 76
    paddy rice culture in
    Southeast Asia, 75
    raised field agriculture,
    Asia, 88
    raised field agriculture,
    Mexico, 87
    shifting cultivation,
    Africa, 78
Traditional knowledge, 25
    ethnobotanical, 89
Vegetable living-mulch
system, 136
Vegetational patterns, 62
Weeds, 173
    allelopathy, 176
    and insect
    populations, 181, 185
    crop competition, 174
    ecology of, 173
    ecological role of, 179
    growth characteristics
    of, 173
management of, 178
Western thought, 22, 25
Western world view, 22